A WIDER VISION

*To my parents, Dick and Mary
They showed me their prayer
and set me free to discover my own*

Gerry Pierse CSSR

A Wider Vision

REFLECTIONS ON GOD, PRAYER AND CHURCH
IN THE LIGHT OF CHRISTIAN MEDITATION

THE COLUMBA PRESS

First published in 1993 by
THE COLUMBA PRESS
93 The Rise, Mount Merrion, Blackrock, Co Dublin

Cover by Bill Bolger
Origination by The Columba Press
Printed in Ireland by Colour Books, Dublin

ISBN 1 85607 064 6

Nothing contrary to faith:
Rt Rev Angel N Lagdameo DD
Bishop of Dumaguete
August 15, 1992

Approved for publication:
Very Rev Fr Ramon Fruto CSSR
Cebu Vice-Provincial
August 1, 1992

Copyright © 1993 Gerry Pierse CSSR

Contents

Preface		7

PART I: A WIDER VISION OF GOD

1	The God Created by Story	12
2	The Triune God	16
3	The God We Depend On	18
4	The Companion God	21
5	The Dependent God	24
6	Fundamentalism and Freedom	27
7	Experiencing God as Gift	30
8	Limiting God	33
9	Our God Image	36
10	Your Accent Gives You Away	39

PART II: CHRISTIAN MEDITATION

11	Christian Meditation	44
	The First Teacher	45
	The Second Teacher	45
	The Teaching	45
	Some essentials of John Main's teaching	46
	John Main OSB (1926-1986)	46
	The Mantra	47
	Choosing your Mantra	48
	Being attentive to the Mantra	48
	The Importance of the Mantra	49
	Silence	50
	The Pernicious Peace	51

	Distractions	51
	Posture	52
	Time	52
	Breathing	53
	Expectations	53
	Meditation Groups	54
	Is Meditation the same as Contemplation?	54
	Is Meditation Christian?	55
	Is Meditation Dangerous?	55
	Technique and Discipline	57
12	The Ego	58
13	Meditation and Washing Socks	61

PART III: A WIDER VISION OF PRAYER

14	Is Meditation Prayer?	66
15	Prayer Beyond Talking	67
16	Prayer Beyond Wanting Anything	71
17	Prayer Beyond Insecurity	74
18	Prayer Beyond the Fear of Prayer	77
19	Prayer Beyond Aligning With Power	81
20	Prayer Beyond Words and Theories	83
21	Prayer Beyond Infringement of Space	85
22	Prayer Beyond Empty Ritual	88
23	Prayer Beyond Compulsion	92
24	Prayer Beyond Asking For Help	95
25	Prayer Beyond Miracles	98
26	Prayer Beyond Frustration	103
27	Prayer Beyond Past and Future: The Mass	107

PART IV: A WIDER VISION OF CHURCH

28	The Centralising Church	110
29	A Church that Reflects the Trinity	114
30	A Wider Vision	118

Preface

Some years ago, in the Philippines where I am a missionary, I heard of a mother whose child was sick. To save the child an operation was necessary and for that she needed the equivalent of about twenty pounds. She had no money, the child had no operation and consequently died. About a week later, the grieving mother was sweeping the house. She noticed a loose floorboard and listlessly lifted it. To her amazement she found a bundle of money equaling about fifty pounds. It had been hidden there by her father who had lost his memory. On finding the money, the anger and the grief of her child's death hit her in a new way. All the time she had had in her house the resource that would have saved her child but, because she did not know of it, she could not use it and so her child died.

I feel that many Christians are in a similar situation. God has revealed a little about the divine to us and because we have not effectively heard it, and have so poorly understood it, we are often in a spiritual death. A Father/Mother God, a Son/Brother God and an indwelling Spirit God, have been revealed to us. But we are poorer for the fact that we have little real awareness of this reality and it's importance for us in our lives.

I was taught catechism as a child, I had a novitiate training, studied for the priesthood and was a priest for about twenty years before I became personally conscious of the significance of the Trinity in my spirituality. For me, God was out there somewhere in a place called heaven and that was it.

One day Cardinal Spellman was called to a man who had collapsed on the streets of New York. He asked the man 'Do you believe in God the Father, the Son and the Holy Ghost?' The man looked at him in anguish and said, 'I'm dying, father! I need forgiveness and you are asking me riddles!' That, unfortunately, is also how I felt about the Trinity for most of my life. For me one person was the whole God, one mode of presence was the whole presence.

Jesus was conscious of the different ways in which God expressed his presence. One day he asked the disciples, 'Who do people say that I am?' He was interested in the perceptions and feelings in the locality, in what we would now call the 'signs of the times'. They answered that some said he was Elijah, others Jeremiah or one of the prophets. Then he asked them, 'Who do *you* say that I am?' He wanted to find out the truth that was *within*. When Peter professed his faith that Jesus was the Christ, the Son of the living God, Jesus replied, 'Blessed are you, Simon Bar-Jonah, because no mere man revealed this to you but the Father who is in heaven.' Truth and knowledge and experience of God come from around us, within us and above us. These are not in competition with one another but complement one another. To have a perception of God that is as balanced and as whole as possible, we must be in touch, as fully as possible, with all of these sources.

Our history tells us that St Patrick brought Christianity to Ireland in 432. The national symbol of Ireland, the shamrock, goes back to the story of Patrick preaching about the Trinity to Laoghaire, the chief of the Irish clans who came to seize the saint for daring to light the Easter fire on the hill of Slane. In trying to explain the triune God, he picked up a shamrock and taught that, as one plant could have three leaves, the one true God had three persons. I believe that the truth that this analogy was trying to express has not yet been grasped. We tend to take one leaf of the shamrock and make it into the whole shamrock.

That is why it has so often been said that Christian renewal must be a contemplative renewal. It is a renewal that comes from our *con-templare*, our living within the same house with God, our living with him who lives within us. Jesus said, 'I have come that you may have life and have it to the full.' It is a waking up to the good news of a God whose Son said, 'Father forgive them, for they know not what they do.' When we are aware of having the Spirit of this God at our centres, we can enjoy life to the full.

There is a powerful image of renewal from the world of computers. It has been found that the rechargeable battery in a portable computer has a memory of its own and a lazy one at that. It remembers the first time it was recharged and will not charge beyond that limit in later charging. If, for example, a battery had a capacity of six hours but it was recharged after one hour in its first use, it will tend to be usable for only one hour in the future. But now there is a programme which brings about a 'deep discharge'. It enables one to work back and recover the lost potential of the battery. Our Christian challenge today is, through contemplative renewal, to do a deep discharge and recapture the full potential of the Spirit, which we may not have been availing of, in our personal and communitarian lives.

My own growth in this conviction has come from my faltering efforts to meditate, to pray by trying to be still and to say a prayer word, a mantra. I had been fumbling at this off and on till I came into contact with the tapes and writings of John Main in 1984 and later with those of Lawrence Freeman, both Benedictines. Their teaching had a simplicity and an authority that appealed to me. As I tried to meditate, to just be present to the mantra, I found myself also becoming more present to the sacred Scriptures, to people, and to liturgical celebrations. Having tried to sit with the Spirit, I found myself going back to seek the Son and in him know the Father. Having tried to sit with the Spirit, I found myself going out into the community with more unhassled availability and compassion. I found myself with a greater love and respect for the institutional Church, even if I do not hesitate to bombard it with filial criticism.

St Paul's image of a husband loving a wife as Christ loves his Church becomes very meaningful for me. Love in marriage does not grow easily. It needs a lot of empathy for the struggle of the partner and a lot of forgiveness of failure. So too Christ loves his Church, even his institutional Church, well aware of, but truly forgiving of its human fraility.

In 1991 I had the privilege of sharing on meditation with groups in Ireland, England, Singapore, the Philippines, Australia and New Zealand. All seemed to be able to resonate with the experience that being with the Spirit in meditation twice daily opens one up to the fullness of the Trinity, the grace of forgiveness, and the richness of the Church in its many expressions. To meditate helps one to stand (Latin *sto*) in the middle (Latin *medio*) between extremes, to be able to accept paradox and grasp wisdom, to be able to say 'Yes' and 'No', not just 'Yes' *or* 'No'!

This book is an effort to share a little of how meditating has helped me come to *A Wider Vision* and a better understanding of the Trinity, prayer, and the Church.

In the first section, *A Wider Vision of God*, I try to share some reflections on what the Trinity means for me. It is not that I know more about this mystery than anybody else, but just that my experience has opened me up to seeing different aspects that may be worth sharing.

These pages will be of little use if the reader is not inspired to set out on, or to continue on, a journey of trinitarian prayer. For that reason there is a whole section on Christian meditation – on how to meditate and on the questions most often asked about meditation.

For those who have never meditated, it may be good to read this section first and begin to practise meditation even as they read a chapter or two daily. Seasoned meditators may also find some help there for the daily new beginning that meditation always entails.

The third section aims at giving *A Wider Vision of Prayer* and the last part *A Wider Vision of Church*.

I would like to express my gratitude especially to three people for their help and encouragement in the preparation of this book: Mrs Carmen Pierse, my sister-in-law, John Wiersma, my parishioner and Seán O Boyle, my publisher.

May your journey be directed towards fullness in the Trinity, symbolised by the shamrock. May it be a journey in which you do not become stuck in any one of the leaves but rather be present to the whole shamrock.

Gerry Pierse C.Ss.R,
Dumaguete City,
6200 Philippines.

A Wider Vision of God

The God Created by Story

During my last home leave I stayed much of the time with my brother, his wife and their two children Nevil and Aoifa. Friends of theirs, who had to go away on a trip, arranged with my brother and his wife that their daughter Kay, Aoifa's classmate, would stay with us. Kay had one problem. She said 'It must be awful to stay in a house where there is a priest. You have to be good all the time.' 'Oh no, it's not like that at all,' Nevil explained, 'You see, because he is a priest, you can be sure he will forgive you all the time.'

This is quite an interesting story. Obviously because of the stories that Kay had heard she believed that priests were rather awful moralistic people. Nevil, on the other hand seemed to have experienced that priests were not like that at all. Many people get their God image from their priest image and so Kay was likely to start off in life with a very different God image to that of Nevil.

It may seem strange to us at first but it is stories that create our world rather than our world that creates stories. If a young person has only heard stories about making money then life will be seen as a money making venture. If the young person has heard stories that approve generosity then generosity will be seen as a goal in life. So, the stories we hear are very important, they give us a vision which affects our feelings, which affects our physical and mental health and our relationships to ourselves, others, the world and God.

One time I was staying in a house and little eight-year-old Ricky came in. He was clearly a precocious kid and I got a sense that I was being set up for a party piece. It went like this:

'Ricky, tell Father what you are going to be when you grow up!'

Ricky: 'Oh, I'm going to be a priest.' Applause!
'Ricky, tell Father what kind of priest you are going to be.'
Ricky: 'Oh, I'm going to be an army chaplain.'
Giggles and laughter!
'And tell Father why you will be an army chaplain.'
Ricky delivers the punch line to the delight of all.
'Because then I will be going from place to place and can have my woman in each place.'

See what I mean about stories? Ricky may go to a Catholic school, receive all of the sacraments, be married or ordained, but he has learned from the stories that he heard in childhood that commitment either in marriage or priesthood is a joke and that women are playthings. Ricky will be what these stories made him.

Everybody has a God and a religion in the sense that there is a central truth around which they tie (Latin *legio*) life together. According to the stories they have heard their god may be money, fame, power, sexual prowess or a myriad of lesser gods. They may believe that they are Christians, but they may not be so at all if other stories are more central to their value systems than the Christian story.

To be a Christian is to be in a world created by the Christian story, the story told in the Bible. As I said already, our stories actually create our world and our God and not the other way around!

The story we find in the Bible is one of great dignity. We are told in the book of Genesis that God created all things good and that he created human beings, male and female, at the pinnacle of his creation. In this simple story we have the basis of ecological concern, human rights advocacy, ecumenism, the theology of liberation, co-responsibility and a host of other affirmations of basic human dignity.

But we also have bad news; from the beginning humans fought back against creaturehood and aspired to be gods. As they sought to displace God and put themselves at the centre they sinned. The Ego was already seeking to be at the centre. In the primitive Old Testament cultures God's vengeance was often seen in the terrible wars and misfortunes inflicted on the people. Yet, even the Old Testament shows a relenting God, a forgiving God, a God of immense patience with human fickleness.

This God so loved the world that he sent his only begotten Son not to judge the world but to bring a message of good news, good news that would enable his people to have life and have it to the full. This message is in every page of the Gospel. Our God is a God that sends his rain and his sunshine on the evil as well as the good. A God whose love and forgiveness cannot be expelled by our lack of love and forgiveness. Jesus tells parables to bring home his message of the Kingdom that is within us, and he is himself a constant parable.

Parables are stories that subvert the shallow views and ways of the world. The line of thinking of the world is that Jesus is good and that therefore he should consort with good people. But he consorts with public sinners and with prostitutes and has the hated Samaritans as the heroes of his stories. The line of thinking of the world is that the sinner is punished but he tells a story about a father who was rejected by his son and yet embraced him on his return ... What a story of forgiveness and love!.

Then Jesus himself became a parable – a subverting story. The wisdom of the world would have God becoming human in splendour and riches but he came in poverty. The wisdom of the world would have God conquer the world but instead Jesus accepted suffering, pain, rejection, crucifixion and death. He joined us in our suffering, giving a fantastic insight into the meaning of consciously embraced suffering itself.

We may not see meaning in our suffering. But if Christ, the Son of God, could not avoid suffering, then suffering must have a meaning and a value that is beyond our capacity to understand.

Jesus tells us to be perfect as the heavenly Father was perfect. What does that mean? That we should be totally sinless? I do not think so! The heavenly Father was perfect by accepting all, by tolerance and patience towards all. The good news is that God loves us no matter what we do, even if we are sinful. When we are aware of this love we can accept our sinfulness in the warmth of that love. We will not need to fear. When we are made to feel bad we have very little choice left except to be bad. When we are made to feel good then we can be good. Jesus showed this for example in the way that he dealt with the woman taken in adultery (Jn 8). He made her feel good, that she was not the only sinner present, before he asked her to be good. In fact, the only way by which we can go from an attitude of fear into one of freedom is through accepting God's incredibly forgiving love.

When Jesus left us physically he sent his Holy Spirit to be with us in a special way till the end of time. He is with us still through that Spirit who, St Paul tells us, dwells in our hearts.

Perhaps one of the great reasons for sadness in today's world is not all the bad news that we hear but the fact that so many people, even religious people, have not heard the Good News, the Good News of our loving and forgiving God.

Unfortunately, like Kay, many people have learned about God in a context of fear. The stories they heard and often the religious role models they saw, gave them the idea of a Policeman God, a kill-joy God, a God to be feared rather than loved.

Prayer is the way in which we relate with reality, the supreme reality, the Triune God. In these pages we will be trying to understand the Triune God a little better and to see how meditation helps us to relate with God in a way that will bring us from fear into the freedom that forgiveness brings.

The Triune God

We do not know very much about God. From the stories we are told in the Old Testament we can deduce a certain amount. We deduce a creator, a liberator, a saving God and so on. What these human analogies really mean is even harder to explain. By the time the New Testament was written down, the revelation that there is one God, and that this God is three persons, was clearly articulated. For example, at the Annunciation we are told that God sent an angel to announce his Son and that Mary conceived by the power of the Spirit. At the Jordan when John baptised Jesus the Holy Spirit was seen in the form of a dove and a voice was heard from Heaven saying 'this is my beloved Son, hear ye him.' In very many places there are explicit references to three modes of the being of God.

If God reveals this information about himself to us, it must be mightily important. Yet, it is only in the last few years that I discovered for myself that it is important. I was a priest for twenty years before I owned, in a real and meaningful way, the revelation that God was one and three. I think I was the poorer on that account.

Let's try now to unpack the idea of the Trinity a little. I will start with a human comparison.

Fifty years ago, I was a little baby and I depended on my parents for everything. Twenty five years ago, I was a newly ordained priest. I went home to visit my parents – who were then about my present age – and enjoyed an adult-to-adult relationship with them as equals. Later, as my parents aged they depended on their children to make most decisions for them.

My mother of fifty years, my mother of twenty five years ago, and my mother in her final days was one and the same person, and yet was not the same person. What was once a relationship of dependence on her became one of equality, and then one of her depending on us. But if I am to remember my mother now I must remember all three. She is and is not all of those three persons.

So too our God is transcendent, he is the God who created us and we depend on him/her for everything. This God is also incarnate. In Jesus Christ, God became human, one of us, and shared our lot. Our God is also immanent and dwells within each one of us. He depends on us to become his hands and feet, his ears and lips, to make him present in the world today.

When my mother was still alive, I was very locked in to the stage of the moment. As a child I experienced her as the one on whom I depended. Later I experienced her companionship on an equal footing. All of those memories were practically forgotten when faced with personality changes brought on by Alzheimer's disease. The problems of the final years made us forget the joys of the earlier phases. After her death we were able to remember her once more in all of the phases.

So too we can as individuals or as Church get locked into one mode of God's revelation as Father, Son or Holy Spirit. To the extent that it is incomplete, to take any one aspect by itself, it is also inaccurate and may mislead us in our response to God.

The name or names that we give to God have far-reaching consequences. Our name for God affects where we perceive God to be, how we pray and how we perceive the Church.

To be in relationship with God – to pray – is to be in relationship with all of the persons of God. If God found it so important to reveal those persons to us, then each one of them must be important and complementary to the others. To neglect any one of these relationships is a misrepresentation of the truth about God and therefore could even be termed heretical. Heresy has sometimes been defined as taking part of the truth and making it into the whole truth.

The God We Depend On

The traditional Catechism of the pre-Vatican Council II days began with the question:

'Who made the World?' and the answer was, 'God'.

Then it asked 'Where is God?' and the answer was, 'God is everywhere but is said principally to be in heaven where he manifests himself to the blessed.'

The message really got home. When I ask patients in hospital or children in school 'Where is God?' the answer is one hundred per cent 'Heaven'.

The problem is that the answer is true. It is a biblically revealed fact that God is our Creator, we locate 'him' in Heaven and we believe that he manifests himself to the blessed who are there with him. But the other problem is that if we take part of the truth and treat it as the whole truth we end up in heresy. By over emphasising the foregoing truth we may ignore the fact that God has also called us and empowered us through his Holy Spirit to be his co-creators and co-redeemers. He also became Emmanuel, God-with-us. He also manifests himself to and in the poor and the oppressed.

The fact that we locate our God and creator 'out there' is reflected in many ways in our lives.

If God is 'out there' truth is 'out there'. Theology becomes a matter of reading fixed, a priori, inflexible principles. The truth is there once and for all and human beings are to conform to that truth or be cast into hell which is 'down there'.

If we depend on a God 'out there', prayer becomes a way of getting to that God. We were told in the past that good prayer was to build an 'altar' to God.

A was to adore the God out there. L was to love him. T was to thank him. These functions took secondary importance usually to the next A, which was to Ask him. Finally, we made (R) Resolutions by which we made a pact to do something heroic for God and felt that we were likewise binding God to do our wills. In this kind of prayer we want God to intervene and solve our problems for us.

We took the 'Our Father in heaven' bit so seriously that we built our Churches with spires pointing into the skies. We offered sacrifices with incense to please the divine sense of smell. (In the Old Testament it was believed that God liked the smell of roast meat and so they made burnt offerings to him). In this perspective the sacramental species were so sacred that they had to be held in vessels of gold or laid on linen cloths and the ordinary Christian could not touch them.

Consistent with this idea we had a Church that was very hierarchical and institutional in nature. Truth came down from above, from God, through his chosen leaders. These had to be listened to and obeyed. The sacramental system was sometimes depicted as a huge tank of grace with seven taps or faucets. The church had control over these outlets. If you wanted to drink of God's grace you had to come to the Church.

Now the problem about this Father-in-heaven picture of God, ourselves and the Church is that much of it is true. But being an incomplete picture it is also false. An attitude to prayer that relates to only one aspect of God is not wrong in itself, it is just inadequate, and it provides an unbalanced spiritual diet.

Fortunately, within this Church there was a mass of 'ignorant laity' not too contaminated by formal theology. Theology told them that God was far away but their *sensus fidelium*, their God-given God-instinct, told them that God was close. So they expressed that closeness in popular devotions. Through mediator saints and a virginal loving mother they softened the austerity of the official line. Through their prayers and devotions, their altars and processions, they were trying to find another way to the loving forgiving God. The church of popular religiosity and devotions, though awry in some ways, testifies to the sense of truth and the desire for balance that is in the heart of the masses of the Christian faithful.

When we talk about the prayer of Being, of be-ing with the God who is Spirit, in Meditation, we are not talking of destroying or replacing specifically Father-centred prayer or eliminating traditional devotions. We are just trying to restore balance. Some meditators have shared how this happened in their lives. Whereas before they would compulsively perform their devotions to their patron saints, after practising meditation for some time they found that they could freely decide to pray to them or not. What seemed to be happening was that as they became experientially aware of the loving God within, there was no more need to be compulsively placating the fearful God outside.

The Companion God

The people of God who belong to the Basic Christian Community in San Fernando, Bukidnon, Philippines, come together regularly to reflect and to pray. They try to read the 'signs of the times' or by looking at their situation and their relationships with one another try in a prayerful way to understand what God is telling them. Their experience is that God continues to be revealed to us every day, if we only have the eyes to see and the ears to hear. Sometime in July of 1987, they became aware that the wholesale logging being carried out in the area would mean that there would be no future for their children. The environment was being systematically destroyed in the name of 'progress'. After much prayer and reflection, they decided that as a community that follows Jesus, they would block the road and stop all logging trucks from going through. Realising that the logging companies had more money, power and pressure than they, their only source of salvation was themselves, and God. One of them was Clarita, and although she was three months pregnant (and was later beaten up by the picket line) she decided to join for the sake of the unborn child she was carrying.

> God came to live among us over two thousand years ago. The good news of today's Gospel is that he continues to come. How happy we would be if we only had the eyes to see and the ears to hear!
> *Bible Diary, January 6, 1992, Claretian Publications, Manila.*

The story goes that one winter evening, as the light was fading, Brother Leo went to the hut of St Francis of Assisi to light a fire for him. It was a cold clear frosty night. Francis looked out over the valley and asked 'Why is there no smoke rising from the houses of the poor?'

'The poor,' Brother Leo explained, 'cannot afford to have fires.'
Francis paused for a little while and then said,

'I cannot take away their cold but I can join them in their suffering, Brother Leo, please don't light the fire.'

This, to me, is the story of incarnation. The story of a God who so loved the world that he sent his only begotten Son not to take away our suffering, but to consciously join us in our suffering so that we may know that in some way it has meaning.

Most of us have little problem in believing that Jesus Christ is God, but we have a lot of problem in accepting that he is human. Most of us have little problem in accepting God's heavenly presence, but a lot of difficulty in accepting that presence in the world. When Christ was on earth this was the pressure that he was always under, the pressure of being human, totally human, while still being divine. In the temptations he was being tempted to use his Godness to become rich, famous and powerful but he resisted these temptations. He was constantly being asked to prove himself. Would it not have been most dramatic if he came down from the cross when asked! One of the big fears of Jesus was that people would follow him for the wrong reason. We see him again and again telling people he healed not to tell anyone. He did not want people to follow him as a magical healer. He wanted people to have faith in God as a God who was truly in-tegrated into this world; a world that is full of suffering, frustration and contradiction.

Our religion is based on the Jesus story and the amazing thing is that there is so little in that story of what we so often associate with religion. One of our expectations in religion is magical relief from suffering, but he himself consciously suffered torture and even death. We look for extraordinary manifestations, but he worked in very simple human ways. As we companion one another in the community he continues to work in very human ways. This is exemplified in the story of the people of Bukidnon with which this chapter began.

Let us look at another modern day intervention of Christ – a modern miracle. Luping had four daughters. Three of them got married early and the fourth, Nora, was fated, rather than chose, to look after her mother. While the family and the culture expected her to do this, she had her own plans and desires. She disappeared without warning one day. When Luping eventually heard that Nora had gone off on a ship with her sailor boyfriend she collapsed, and she recovered only twenty four hours later in hospital. She then swore that her ungrateful daughter would never again set foot in her house.

In that neighbourhood, the Basic Christian Community had Bible sharing in different houses. Two weeks after her coming out of hospital, Luping was hosting the affair. The reading was the story of the forgiving father from Luke Chapter fifteen. They reflected on the passage. (One man asked if this bad story could be removed from the Bible; wasn't it clear that the father should not have forgiven that son!) The different parents shared about how they had suffered because of the ingratitude of their children. They found themselves caught between the values of the story and the values they found in their own culture. They gradually accepted that their traditional concepts needed revision. Luping said nothing during the discussion but she was deeply shaken. She had suffered so little in comparison to some of the other parents and compared especially to the father in the Gospel. Gradually her attitude began to change. On my last visit to Luping, Nora was there again.

This is how the forgiving companion God is reincarnated in our midst in our times.

THE COMPANION GOD 23

The Dependent God

A young Jewish couple were very poor, but the poverty that hurt them most was the fact that they were childless. They went to the rabbi and asked him to say the prayer that would give them a child.

'Certainly, I will say the prayer,' the rabbi said, 'but first you must give me $500.00.'

'Oh, Rabbi,' they said, 'we have never even seen that much money!'

The rabbi would not yield and they went away disconcerted. They returned next day and said:

'Rabbi, we know that you are a good man and we think that maybe you did not get the picture correctly yesterday. We are very poor. We desire a child so much. Can you help us by saying the prayer that will give us a child? If you want us to pay, name some amount that we can afford.'

'Certainly, I will say the prayer,' the rabbi said, 'but first you must give me $500.00.'

'Oh rabbi,' they said, 'we told you we have never even seen that much money!'

The rabbi would not yield and they went away with sad faces.

They returned next day looking radiant with joy and said:

'Rabbi, Rabbi, we have made a decision.'

'What is it?' he asked.

'We have,' they said, 'decided to say the prayer ourselves!' And the rabbi beamed with joy and said 'Amen.'

There is a fundamental truth about God in this story. God does not want us to be asking him to do what we should be doing ourselves. He loves us too much to answer our often very silly prayers. He dwells in us. There is a power in us, a power that we are often fearful to use. He puts us on the spot so that we will recognise and use our power. He wants us to overcome our fear and use our talents. He empowers us and dis-empowers himself. He creates the situation where his power can only be seen when we use the power and the gifts that he has given us, especially the power to forgive ourselves and others. This God is the dependent God, the Holy Spirit God.

The fearful virgin Mary opened herself to this power when she said, 'Behold the handmaid of the Lord, be it done to me according to your word.' Then the power of the Holy Spirit overshadowed her and she conceived and went off to help her cousin Elizabeth.

This God was manifested especially in the experience of Resurrection and Ascension in the early Church. The event of the Resurrection established that God truly walked with human kind. However, this God-with-us was not easily recognised. Even his closest disciples did not recognise him until there was first some engagement, some involvement with him. He ascended to heaven without leaving exact and precise instructions on anything. Rather he empowered his people to act on his behalf and to forgive each other as he did. This demanded knowledge and courage.

This direct empowerment by knowledge and courage was very apparent in the early Church as recorded in the Acts of the Apostles and in the Epistles. Even more important, and seldom emphasised, is the fact that the first gift of the Spirit is the gift of forgiveness. 'I give you the Holy Spirit ... When you forgive sins they are forgiven.' He was not here talking primarily of a sacramental power given to priests but of the power in each person to bind or release themselves and others in the community. There can be no freedom and no community where there is no forgiveness. We are equally challenged today to be present to that power, that gift, that freedom, and to let it work so as to let God work through us in building community.

The power of the Holy Spirit is a bit like how rabbits live. They live within the hedge or fence and will not come out when there is hustle and bustle. They only come out when there is silence and stillness. In a similar way, the dependent God does not push himself. He is always there, powerfully there, but he needs space and time so that he may open Heaven's gate for us. That is what we do when we meditate. We are silent, we say the mantra, we take the focus of attention off ourselves so that the Spirit can be. He can then come out from our very depths and he gives light to our darkness and courage in our fears. He shows us where he wants us to go and what he depends on us to do.

The Spirit within us is more intimate to us than we are to ourselves. The Spirit within us should be our first place of prayer. It should be our first but not our only place of prayer. Our towns have water supplies. Most of these originate in a lake or reservoir. The water is then pumped to a tank on an elevated place. After that it is allowed to flow by gravity, bringing this essential for living, to the faucets or taps in our homes, gardens and places of work. The water in the lake, tank or faucet is the same water. When on a picnic you may go to the lake and take a pail full of water directly from it. You may rush, as the fire engine does, to get water from the tank. Normally, however, for your day to day usage, you will turn on your tap or faucet right there in your home if you need water.

So, too, we can and should turn to God our Father-Mother-Creator on whom we depend for all. This is the main emphasis, but not the exclusive one, when we go to church and worship. When we read the Scriptures and reflect on the words and deeds of Jesus Christ, we are principally but not exclusively growing in relationship with Jesus, the companion God, who became one of us.

But just as we turn to the faucet within our house for our day-to-day water needs, so too, our day-to-day relationship with God will principally be with the Spirit who dwells within us and who helps us to express him in and to the world.

Fundamentalism and Freedom

As a child I used to enjoy cowboy and Indian movies. Very often we would see the wagon train crossing the prairie when suddenly Indians would appear whooping over the horizon. The wagon train would immediately form a circle. From within that circle the cowboys would shoot at the circling Indians and the Indians rained arrows on the cowboys.

This is a good image of what it is to be conservative. When under threat we form ourselves into a protective circle to defend ourselves against a real or imaginary enemy. This can be very good and very necessary. However, if we are in constant fear of the enemy we may never move out of the circle. If we do not take the risk of moving out of the circle we cannot make progress. Those who tend to stay more in the circle are considered conservative. Those who tend to move out, sometimes taking great risks, or sometimes just unaware of the danger, are considered enterprising and/or liberal. Neither group has a monopoly of wisdom or of foolishness.

Today's world is one of great pressure and often of great fear. The advances of modern technology are mind boggling. This can be frightening for those who were at one time at the top professionally and now find themselves computer illiterates. Now there is much greater freedom of thought, topics that were once considered to be taboo are being discussed openly and endlessly. In the past too there was greater certitude of religious truth. An authoritative church told us what was right and wrong and the majority accepted this without question. Today there are few issues that are clearly black or white. Gray areas emerge and the questioning of what was once considered sacred can be very threatening for many.

There can be moments in all of our lives when we are faced with terrifying questions. I remember when one of the Brothers in our community died. As is the custom in the Philippines, he was embalmed and laid out in the church. After a few days the undertaker came to top up the embalming. I was asked to assist him. As I saw the partly exposed body I was overcome with radical doubt. Could this be the temple of the Holy Spirit? Is all this talk about Resurrection a big hoax?

Another time a ship sank in which a few people I knew perished. I gathered with some friends of theirs to celebrate the Mass. It was the fourth Canon of the Mass that was prepared.

'Father, we acknowledge your greatness;
all your actions show your wisdom and love'

I slammed the missal closed and said 'Come off it, Lord, who are you trying to fool.' I could not say those words sincerely then because I did not feel, at the time, that they were true.

Life does confront us with situations where individuals or groups can see no meaning in what is going on and feel totally powerless. Many years ago Eric Fromm pointed out that the popularity of Mickey Mouse cartoons went back to the sense of powerlessness that people felt. The situation in the cartoon would often have little Mickey being harassed by a huge cat. Then he would emerge from his mouse hole and give the cat a prick of a needle in the posterior. We would all enjoy it because we identified with the powerlessness and oppression that Mickey was suffering and also with his moment of triumph.

This kind of situation is faced in all walks of life. It is met in different ways one of which is Fundamentalism. Fundamentalism is the reduction of reality to some apparently manageable norms. It becomes irrelevant whether these norms represent reality correctly or not. Once these norms are adopted they must be defended and no deviation is tolerated. If I establish that by following these norms I am right I will have to say, at least by implication, that everyone else is wrong. In the time of Christ the Scribes and Pharisees had reduced life to following laws. Thus when the woman taken in adultery was brought before Christ (John 8) they said, 'We have a law and according to that law she should be stoned to death.' Jesus, however, saw the situation more widely and upheld the dignity of the woman. Today's fundamentalists tend to take a literal interpretation of Scripture as the norm of truth. When they are 'born again' they alone have the truth and they tend to impose it on others. They wear blinkers over their eyes so that they cannot see the truth that comes from any other source.

Our God is indeed exasperating. He is always bigger than our boxes. He can work through an oversexed autocrat like King David as well as well as through a boastful bungling Peter. He was always open to saint and sinner alike. In fact, he considered self-righteousness the worst sin.

In my home town there was a story about a lawyer who had a very weak case to present in court. One day his client said to him, 'You know, I have some nice fat geese at home. Would it help if I gave one to the Judge?'

'Are you mad,' answered the lawyer. 'Do you not know that this is the most upright and self-righteous judge in the whole coun try!'

They went into court and the presentation went even worse than expected. However, the judgment was given in their favour and the lawyer was totally amazed. Later, the client said with a smile,'See how the goose worked!'

'Don't tell me,' said the lawyer, 'that you sent the goose!'

'I did,' said the client, 'but I sent it in the other fellow's name.'

There is great irony in this story because it is the righteousness of the judge that made him blind to the truth and therefore act unjustly.

Boxing ourselves into any kind of box or narrow category is very dangerous. It makes no difference whether this box is one of righteousness, of a God concept, or of a way of prayer. God is more and bigger than all of them.

Experiencing God as Gift

To begin this reflection I would like you to recall times that you have felt happy. Maybe there was the time when you were accepted for some desired position or there were times when you won in a competition or passed an exam; they were times of achievement.

There was also another kind of happiness called joy. This came at times when you discovered that you were loved, that some one really cared, when you got a gift that you did not expect. This kind of happiness is in the experience of gift and in the lack of expectation.

All happiness is one of these two; associated with achievement, things that we worked hard for, or associated with gift, happenings or things that we never expected and never felt that we deserved.

Let us look at them more closely. The first is associated with effort, planning, control, expectations, competition, rivalry, jealousy, never having enough. Life's load is found here. Disappointment, jealousy, failures, unfulfilled expectations also go with this kind of happiness. Joy, the second kind of happiness, is associated with the experience of gift – of something coming to us freely from outside. It is something that we never felt we deserved, something that makes us feel that we have more than we need, that we would like to share. These gifts challenge us to respond, to let go and trust. Very little of life's load is found in this area.

In the parable of the loving father and the prodigal son we see the difference between earning and accepting loving forgiveness. The older son thought that the father's love was to be earned and he was resentful of the father's largess towards his younger brother. The second son, however, experienced love as a gift when he never expected to be forgiven.

People tend to approach life more predominantly from one of these points of view. There are some who see life as a challenge in which they must achieve, in which they must fulfil their own expectations or the expectations that their parents, their culture, and others have imposed upon them. There are others who accept life as a gift, to be enjoyed, to be responded to, to be shared.

People in the first group have high expectations and believe that they must achieve. Sometimes their expectations are unreal. Because the achievement has to be theirs, other people are a threat to them – rivals to be feared rather than friends to be relished. They are never secure, never satisfied.

The people who see life as a gift are always grateful, never threatened. Jose from Banago comes to visit me from time to time. He is a beggar, but a very different kind of beggar. There is a dignity and joy in him that I seldom find in the ordinary population. One day he told me his story. After the Second World War he used to load cargo on the ships at his native Banago. One day he stepped on some rusty nails and fell down, 'But I was very lucky, I fell on the edge of the pier. I could easily have fallen into the water and been drowned. I developed infection in both legs. I was taken to hospital but again I was very lucky. Only one of my legs was amputated. For some time they thought they would have to take the two. Then a kind benefactor gave me a pair of crutches which I had for five years till they were stolen as I slept one night on the steps of the church. But then a kind passerby gave me a pair of slippers that I tied to my knees and I used them to crawl with for five years till I was given another set of crutches. God has always sent someone to help me in need.' Here was a Mystic. A man from whom no one could take away his joy, who always saw the gift and was able to relish it.

Many people try to bear life's problems with strength and

determination. Jose faced it with weakness and flexibility. When we realise that we are weak we will not be surprised that God's plan for us is different from what we planned. We will be able to let go to it. When we are accepting we will be able to flow joyfully with the current rather than fighting it and the whole world.

There was an old man who had only three valued possessions left to him, his son, his little house and the pig that was caged under the house. One night he heard a scuffle and when he went down the cage was open and the pig was gone. People said 'Bad Luck,' but the old man said, 'Good luck, bad luck, how do I know?' A few nights later he heard another scuffle under the house. When he went down he found that the pig had returned with five wild pigs. He closed the cage. People said 'Good luck,' but the old man said 'Good luck, bad luck, how do I know?' They slaughtered one of the wild pigs to celebrate but during it one of the pigs gored his son and broke his leg. People said 'Bad luck' but the old man said 'Good luck, bad luck, how do I know?' A short time later there was a state of emergency in the nation and the military came conscripting all of the young men for the war. When they saw the young man's leg they would not take him. When asked about it the old man said 'Bad luck, good luck, how do I know?'

Openness to giftedness is the secret of joy. It is also something to which prayer should be bringing us. Unfortunately, most of us when we pray already know what God should do and so we cannot appreciate the gift when we receive it. We are like the man who prayed for a butterfly. He got a caterpillar and threw the dirty thing out the window. As he complained about how God had not answered his prayer the caterpillar had developed into a beautiful butterfly outside his window.

Limiting God

Lately, I spoke to a young man who had difficulty in relating to his parents. He said that any time he tried to talk to them they would first give him a lecture and set up rules for the conversation. He found himself unfree and frustrated even before he started. It is clear that the parents were insecure in their dealing with their son. Because they were afraid to be open they set up boundaries.

We can be the same with God. Because we are afraid to let him be God we set up parameters on how he should act. I once had a retreatant whose mother had recently been diagnosed as having terminal cancer. At the outset he said 'Do not ask me to pray. I have only one prayer "Save Mom." If God is not interested in doing that I am not interested in God.'

This is a very explicit example of something that we are doing implicitly all the time. All the time we are setting parameters to the response that God will make to our prayers. We are laying down criteria for God, telling him exactly how he is to answer our prayers. We are setting limits to his action.

If we reflect on this we see how silly it is. We are putting our puny intelligence, desires and judgments before the wisdom, goodness and justice of God. We are telling him that we do not trust him to do what is best because we know what is best. This compulsion to control God comes out of our insecurity and fear. When we are insecure and fearful we cannot let go and trust anyone even God. We are tunnel visioned. We see only one way, our way, and feel rejected and threatened at any other solution. Fortunately, God loves us too much to indulge all of our silly prayers.

If you were to ask me what is the most basic problem of most human beings I would say that it is insecurity. What causes us to tell lies, to cheat, to steal? More than anything else it is insecurity, it is fear. We can have fear in facing ourselves, in facing others, in facing God.

We usually have the same pattern in facing ourselves, others and God. If we are fearful we try to defend ourselves and to justify our deeds. If we are free we do not have to, we can be transparent. Neither do we have to blow about our achievements nor hide our limits. We do not have to be always competing with others. We do not have to be telling God what he should be doing. We can let go of our enslavements and addictions.

God's centre is everywhere but his circumference is nowhere. We can find God at our own centre but we can never put a limit to how far he extends. So too we can never put a limit to his wisdom and generosity. There is a deep fear within us that stops us from trusting God or others. I once heard a priest tell about how he used to visit his aged mother when she was sick with arthritis. He would take her up in his arms and bring her downstairs. Descending the stairs she would run her hand along the bannister and sometimes squeeze it tightly. When this happened he would say to her, 'Mom, why are you holding the bannister?' She would answer, 'Because I am afraid'. He would answer. 'I understand that you are afraid but, you know, if you continue to cling to the bannister we will not be able to move up or move down.'

Letting go and letting God be is the secret of freedom. The test of true Spiritual growth is that one is becoming more able to let go to the Lord in full confidence. In the gospel Jesus sent out the disciples and he told them bring nothing. True richness is not in possessing but in being able to let go of all attachments.

There was a rich man who had a dream that if he went outside the town he would find a poor man under the mango tree who would give him a great treasure. He drove out and, sure enough, there was a man sleeping under the tree. He roused him and told him of his dream. 'Yes,' said the poor man, 'maybe this is what you are talking about,' and he took out a diamond as big as his fist. 'How much? How much do you want for it?' asked the rich man.

'Nothing,' said the poor man. 'If you think it will make you happy, take it away.'

The rich man returned home thrilled at his find and laughing at the poor old fool who could give away a diamond worth millions. However, that night he could not sleep and also the next. On the third morning he returned to the poor man and said, 'Please take back the diamond, but could you give me the real treasure?'

'What is that,' asked the poor man and the other answered, 'The real treasure is the ability to give away the treasure.' Real freedom is to be able to be with God without wanting to control and own him.

Our God Image

(a) There have been periods in my life in which I have experienced great sexual turmoil and even failure to live up to my ideals. (b) There have been periods of my life in which I felt very much at home with my sexuality and experienced a minimum of turmoil. I believe now that I was no less a good person and no less lovable to God in the periods described in situation (a) as I was in those described in situation (b).

I only became conscious of the conviction described above sometime recently. If I were a Spanish or Italian mystic I would probably say that I had a vision, that Our Lord or Our Lady had spoken to me. I believe that it was a greater miracle than the lame walking, the deaf hearing or the sun dancing in the sky. It was basically the miracle of the changing of my God image. When my God image changed my self image changed and, incidentally, my behaviour also changed. It is a miracle that is necessary in the lives of most people.

We are brought up to believe that we are lovable only if we are good. As children we were told 'that's a good boy' when we did something that was considered good and 'that's a bad boy' when we did something that was considered bad. Quickly we began to equate doing good with being good and doing bad with being bad. So if we did not live up to the criterion of good that others set up for us, and soon that we set up for ourselves, we were bad people. Often these ideas were reinforced by the idea that Holy God would be angry with us or that the priest would take us away. If we did bad we were not liked, we did not like ourselves and most certainly God would not like us. We had to be holy for God to like us.

We prayed 'Lord, make us pure and holy'. Purity was equated

with holiness. It was also equated with sexual purity. Sexual purity was equated with a denial of much that had to do with the normal process of growing up as a boy or a girl.

Much of the negative attitude towards the church and the identification of the church with women goes back to puberty. The young lad has felt the stirrings in his loins. He wants to see and feel and experience. He may have masturbated. He has been told that this is a most terrible sin and it would be another worse mortal sin if he was to go to Holy Communion in that state. Sunday morning comes and the rest of the family march up to Holy Communion. What is he to do? If he does not join them they will suspect his dreadful depravity! If he joins them he commits a terrible sacrilege! So he dodges the whole embarrassing thing as soon and as often as he can. The Church and the God it represents puts him in a bind that he cannot endure. He may get away from the external embarrassment but the internal battering continues. He is being told from within, 'I am not a good person.'

Women tend to get caught in this trap in later life. A woman of high profile socially or the head of several religious organisations finds herself masturbating. This is awful! It is not actually so much that it is awful but that she cannot stand the feeling that she is herself a fraud and that she may be found out to be one. There is a constant tension about her 'sin' and a determined effort to get rid of it. This determination creates new tensions which cause the compulsive behaviour to continue. This behaviour becomes the measuring rod of goodness. An internal voice keeps on saying 'If only I can overcome this 'sin' I will be a good person in my own eyes, in the eyes of others, and especially in God's eyes.'

What a waste there is in all of this anxiety! God does not see things in this way at all. God loves us whether we are 'good' or 'bad.' He sends his rain and his sunshine and his love on the 'good' and the 'bad' alike. He loves the prodigal son as well as the dutiful one.

God's estimation of good and bad is very different from ours. He sees the heart and the history of each one. He sees the influences that have formed each one and the pressures that bear on them.

He calls us to be free and faithful in the context of these pressures.

There can be terrible pride in wanting to be perfect. It is wanting to be like God. It is not accepting how God made us. He made us human and loves us as human. Humility is to accept the human imperfect state. This is what scrupulous persons cannot do. Their problem is not just fear of not being accepted by God. It is a problem of not being accepted and loved by themselves. It's Okay to be not Okay. I'm not Okay and that is Okay. St Paul talks about the thorn in the flesh. The constant temptation that kept him from being proud. One of the worst things that could happen to us would be to think that we are perfect.

To know this loving God we need space. Very often because of our poor image of ourselves we enter into frenetic activity especially at the time of prayer. Because of our poor image of ourselves before God we strive to be seen to do the right thing, to be heard saying the right words. But this only keeps God at a distance. This will not enable us to experience him. He is not in the storm but in the gentle breeze.

If in prayer we just be with him who is and experience his unthreatening love we will gradually get over our restlessness.

In that silent being we will come to know him and to know ourselves and become free to be our true selves. When this happens our compulsive behaviour will also tend to change. This is brought out in the story of the man with the stutter that I told already in *Silence into Service*.

This man had a terrible stutter. He had not spoken straight since birth. One day he found himself on a bus without money. To get out of his predicament he decided to exaggerate the stutter when the conductor came along, hoping that he would pass him by in exasperation. To his amazement when the conductor came along he spoke perfectly.

When we accept our imperfect selves in the presence of the God who was the Father of the prodigal son, we too will be amazed at how our behaviour and our lives will change.

Your Accent Gives You Away

After a little while, those who were standing there approached Peter and said to him, 'Of course you are one of the Galileans; Your accent gives you away.' Peter began to justify himself with curses and oaths protesting that he did not know that man. Just then a cock crowed. (Mt 26:73-74)

When I am in Ireland everybody knows that I am Irish. When I am in the Philippines everyone presumes that I am American. But if I am in England or Australia people do not know where I come from until I open my mouth and then my accent gives me away.

While there may sometimes be genuine reasons for concealing where one comes from, it is generally a sign of fear or insecurity to do so. Some people pick up accents very easily but to deliberately change ones accent could also be a sign of insecurity. If one is living in a different culture a certain amount will always wash off and there will be a natural leveling out. There is always some place where ones accent is found amusing or difficult to understand. For others it will be intriguing. To balanced persons this will be a source of amusement and intrigue for themselves.

Everybody also has a theological accent and inevitably we give ourselves away. This is how it should be. Different people, or the same person at different times, are more tuned in to the Father, the Son or the Holy Spirit and to the type of ministry and image of Church that goes with that Person. This is a product of their faith history, of the theology they have imbibed and of the experiences that they have reflected on. Their accent expresses who they are now, theologically speaking. Consequently, there are some people who are not at ease when in institutional garb or clerical dress while others can't bring themselves not to dress that way. The first are uncomfortable with the institution and the others are uncomfortable without it. One could be as unfree as the other.

At one retreat I attended there was a priest who was always dressed in a Roman collar and an immaculate black suit. Was he any less free than Fr T-Shirt Tom who could not put on vestments for a formal liturgy?

Rose decides that she would like to go to a preached structured retreat this year, while Nora wants to take the Enneagram with a group and Tessie wants a retreat in silence. All of these accents show where the person is at the moment. Rose feels that she needs input from above, Nora feels that God will speak to her better right now through interacting with the group and Tessie believes that she will hear him in silence. Their choices reflect the mode of God that they are most tuned to right now. Next year all three may have changed position.

Who amongst us can say which is the most dignified or pleasing to God: a liturgy celebrated crosslegged in shortpants sitting on a cell floor with prisoners celebrating the news of their release; a liturgy celebrated under a single naked bulb hanging from a branch of a tree with the community, including their dogs and chickens, joining in; and a liturgy in a chandeliered Cathedral with golden chalices, ministers in vestments and angelic choirs.

Ministry reflects where we are at in our own personal spiritual journeys. Tom visits the sick and feels that he must administer sacraments and blessings. John finds that reading scripture and praying with a patient is very healing. Tim finds that a genuine attentive presence to a patient is a genuine spiritual encounter and that there is no need for explicit religious paraphernalia.

It is as important to be aware of one's own accent as it is to be aware of that of another. If Tom and John and Tim are not aware of their own accents, they may find themselves unable to read the need of the patient who does not speak with their accent, or imposing their version of ministry on someone who is not able to understand it or to be comforted in receiving it.

When reflecting on accents above we saw that there is no need to deny one's accent. Likewise it would be rather silly not to accept that others have different accents. It would be self-defeating to refuse to try to communicate with those who have a different accent or to insist that they change to yours. It shows great insecurity if all that one can do is to throw scorn on the accent of another.

Two sisters from the same congregation arrived at one of my retreats. One was in the traditional habit and the other in a track suit. 'Are you still a sister?' asked the former novice mistress. 'Yes, I am,' came the bitter reply, 'and you are still the Sergeant Major.'

There is a need for genuine respect for the religious accent of the other person. (Of course, the person may have a real speech defect and not just an accent. That's another matter both in speech and in theology.) One needs to make a great effort to hear it accurately and to respond to it positively because we so easily write people off when we get just the first whiff of their accents.

Christian Meditation

Christian Meditation

Through Christian meditation we 'be at home' with God who is at our very centres.

Our relationship with the transcendent God is generally mediated through devotions and sacraments.

Our relationship with the incarnate God is generally mediated through involvement and action with people and the world. In these relationships we can often be 'Marthas,' busy about many things.

Our relationship with the immanent God is without any mediation, it is direct. This is what mystical means, it is entirely simple and uncomplicated. It is to be a 'Mary', to choose the one thing necessary, to be hospitable to the Lord, to be at home with him, and then to move out into the world from that relationship.

In this section I would like to set out briefly the basics of Christian meditation as gleaned from the teachings of John Main and from my own experience.

In his teaching there is only one thing that is essential; to continue to try to say your prayer-word or mantra during the whole of two daily meditation periods. Everything else is just helpful.

The basic idea of Christian meditation is contrary to our whole culture which is geared towards immediate success. 'Transcendental' and some other kinds of meditation promise 'immediate first time achievement of transcendence.' Christian meditation makes no such promise.

In meditation we just try to be present here and now, we try to let go of all that is not God so as to be at home with God in a totally uncluttered and uncontrived way.

One of the great obstacles to prayer is our obsession with success. In all walks of life we are under pressure to be successful. We tend to be the same in prayer and we tend to judge our success or failure by the presence or absence of distractions. But we do not pray to be successful. We pray to be faithful! The more often we come back to our mantra in fidelity, after being distracted, the more faithful we have been.

This vital truth is summed up beautifully by T.S.Elliot, in East Coker when he says,
For us, there is only the trying.
The rest is not our business.

The First Teacher
The first Teacher is God, Father/Mother, Son and Holy Spirit. To learn to pray is to be in reverence before God our Creator.

To learn to pray is to be a disciple of Jesus who taught us prayer by word and example.

To learn to pray is to be tuned in to the Spirit who is for ever praying within us crying 'Abba Father'.

The Second Teacher
The next teacher is John Main OSB who has been my inspiration and guide. Through following him we can become open to the first teacher.

The Teaching
'You just sit still and upright. Close your eyes lightly. Sit relaxed but alert. Silently, interiorly begin to say a single word. We recommend the prayer-phrase 'MARANATHA'. Recite it as four syllables of equal length. Listen to it as you say it gently but continuously. Do not think or imagine anything - spiritual or otherwise. If thoughts and images come, these are distractions at the time of meditation, so keep returning to simply saying the word. Meditate each morning and evening for between twenty and thirty minutes.'

Some essentials of John Main's teaching
Modern people are very alienated within themselves. Through meditation they take the focus of attention off themselves and become restored to their own centres.

At their own centres they become present to the prayer of the Holy Spirit that dwells within each one of us.

This puts them in touch with great energy and leads to fullness of life.

Meditation can only be learned by doing it. Talking or reading about it too much can delay the process. Only personal experience will convince one of the validity of the claims made about meditation.

John Main, OSB (1926-1986)
For John Main, being rooted was very important. His roots were in Ballinskelligs, Co Kerry, Ireland, though he grew up mostly in London, England. He was born in 1926 and baptised Douglas. Later as a Benedictine he took the religious name, John, by which he is remembered now.

In 1954, as a young lawyer, he joined the British colonial service in Malaya. One day he was sent to deliver a picture to an Indian monk, the Swami Satyananda. This holy man was to have a big influence on his life. They began to talk about spiritual things and the Swami undertook to teach John how to pray by saying a mantra, or prayer word.

In 1956 he returned to teach Law at Trinity College in Dublin. If, during this period, he told any of his priest friends about his meditation they treated it with suspicion or hostility. After the death of a nephew he reflected more deeply on life and decided to join the monastery of the Benedictines in Ealing, England. As a novice he found it very hard to obey his novice master who told him to give up meditation. He was learning to 'let go' of attachments, even spiritual attachments, and this was to become essential to his future teaching.

In 1970 he went to the United States where he was headmaster for five years of the Benedictine School in Washington,D.C.. At this time he had a chance encounter with the writings of Augustine Baker, a 17th century English monk, which led him back to John Cassian, the 4th century teacher of St Benedict. There he discovered the same practice of meditation with the mantra as he had learned from the Swami in Malaya. He was inspired by this to study further into the Christian tradition and found more and more proof of the tradition for meditation that went back to apostolic times. He resumed his life of meditation three times a day and began to teach Christian Meditation to others. One of his disciples at this time was a young Englishman, Lawrence Freeman.

He returned to Ealing and started a lay community of meditators. From then until his death in December 1982, his energies were directed into teaching meditation and forming communities of meditators.

The Mantra

The mantra is a prayer-phrase or prayer-word that is repeated continually subvocally from the beginning to the end of each meditation. The 'man' comes from a word meaning 'mind' and the 'tra' from a word meaning 'to cleanse.' The mantra is seen as a way of cleansing the mind. It cleanses it especially of negative thoughts and imaginings. Thoughts and imaginings, especially about God are, in this perspective, limiting. No thought can grasp the whole of God and so to have thoughts is to be boxed into something limited. We imagine what is absent, we BE with what is present. If I imagine God in prayer I am actually making him far away. The mantra helps to make the space in which to be present totally to the here and now.

The mantra is used for prayer in many traditions. According to John Cassian, who wrote in the fourth century, it's use in Christian prayer goes back to the time of the Apostles. It is probable that the 'Our Father' originated in prayer phrases that were recited by the earliest Christians. The 'Cloud of the Unknowing', written in England in the fourteenth century, tells of praying by constantly repeating the 'one little word.' It talks of the poverty of the single verse.

Choosing your Mantra
The matter of choosing one's mantra, and of then sticking to it, is of considerable importance. This is best done with the help of a director or teacher. The idea of being given your prayer word by a teacher is strong in many traditions. This can be unhealthy to the extent that it might give the impression that the mantra or the teacher has some sort of magical power. But it is important to get a mantra that suits you. Some people will find that after a little trial and error a word seems to come to them. That may be a sign that this is the word for them.

It is generally recommended that 'the word' chosen has a richness from one's tradition. It is also preferably a word that does not have meaning in one's own language. It should also have open vowel sounds.

John Main chose the word MA-RA-NA-THA as fulfilling all of these. It is an Aramaic word which means 'Come Lord'. It is an ancient Christian prayer and the very last word of the Bible. I would recommend that anyone starting to meditate, who has no reason for choosing otherwise should use the word MARANATHA. If a person has been using another word in another tradition and comes to Christian meditation they should keep their mantra if they are happy with it. If one is not happy with one's word or with Maranatha there are many words that are often used, such as Jesu, Yahweh.

I would compare choosing ones mantra to dating. A boy or girl will normally give some consideration to whom they will date. For the date to be sincere there will be an openness to the possibility that this will lead to a more permanent relationship.

Flirting simultaneously with different partners because they appear attractive will make depth of relationship impossible. However, a time may come when incompatibility is obvious and the only sensible thing to do is to terminate the relationship.

Being Attentive to The Mantra
If we want to progress in anything, we make progress through attention. If we want to learn to type or drive, or to use a computer,

we have to concentrate and give our whole attention to what we're doing. While we're still learning, we have to do a lot of concentrating, but after some time, when we've given it attention, it becomes second nature to us.

This is also true in prayer, we have to be alert and attentive when we pray. St Teresa said that if we attend to the 'Our Father', just attend to it, listen to it in our hearts, it can bring us to the highest form of virtue. But very often we don't give attention to what we do, and that applies especially to our prayer. We neither say what we mean nor mean what we say.

When we meditate, we give attention to silence. When we come to being with the Lord, it's not just enough to be in a sleepy inattentive way. When we are asleep we are not conscious or attentive. We need to be focused and to be attentive. This is exactly what the mantra does, as we keep coming back to it, no matter how often we get distracted, we are being with the centre of reality and not with theories, words or ideas about reality. And every time we come back to being attentively there, we are getting deeper into that reality. As we grow into the depth of reality, we are also growing into unity because that is where unity is.

It may be well to note also that meditation is not self hypnosis. In hypnosis there is suggestion. In meditation there is only attentiveness to reality without any effort to control or direct it.

The Importance of The Mantra

In the Christian tradition of meditation itself there are different understandings of the importance of the mantra. In some traditions or teachings the mantra is a help to coming into silence or into God's presence. If one believes that one has arrived at such a state the mantra is no longer needed.

In the teaching of John Main the mantra is always necessary. According to him if one becomes aware of being in silence one has already lost it, so the best thing to do is to again take the focus off oneself by saying the mantra. If one tries to stay with the silence this could be a focusing on self, a celebration of success in having achieved 'silence' or presence. For John Main the surest way is the way of radical poverty. Let go of the 'silence', return to the mantra and God will bring us back into true silence if and when he wishes. It is precisely on this point of emphasis on the mantra that the teaching of John Main differs somewhat from centring prayer.

Silence

To meditate we try to be silent. There are three kinds of silence, that of the body, the mind and of the emotions.

We are growing more conscious today of the connection between the physical, the psychological and the spiritual. If we are to be silent within, we need to be physically silent. When we begin to meditate we will often experience restlessness, a wanting to get away from it. Sometimes we will almost feel our feet walking out the door or that the body has gotten up and left before realising that it has happened.

The next kind of silence is silence of the mind. As we try to be still we find that we have several theatres within our heads all showing their own internal movies. We may be reliving the past or enjoying fantasies about the future. One Indian writer said that the mind is like a tree inhabited by monkeys. They keep jumping from branch to branch and chattering at each other. It takes great patience with ourselves to calm this turmoil and to stop the movies and the monkeys.

The third kind of silence is an emotional one. As we try to let go and become still we may become aware of tension or unrest within. Very often this is due to fear, anger or resentment which we may or may not be able to name. Psychological processing would say that these feelings should be brought out into open consciousness and faced. This is often very helpful. Flight from them or repression are not helpful. By the silence of meditation we deal with these emotions in another way. We just stand our ground. We are not intimidated by them, we just sit still. As we do this they weaken. Bit by bit this letting go, this tenacity bears fruit in wholeness, integration and calmness.

The Pernicious Peace
According to the Fathers of the desert, one of the great enemies of meditation is the pernicious peace, the *pax perniciosa*. It is also the most subtle. It is a state of mental vacuity or absorption during the time of prayer that could easily be mistaken for prayer. It could be a pious absorption or an absorption in one of our fantasies or distractions and it is discovered almost with regret. You have come to a *modus vivendi* with the distraction, almost come into friendship with it. You suddenly realise that you have been 'out' for ten minutes and you are somewhat glad about it. The lazy part within you has enjoyed a break from the discipline of the Mantra. Or you may wonder if you were really into *silence* and really absorbed in the Lord. Maybe I'm succeeding at last!

The tradition here is very clear. As soon as this state is discovered, whether it is a state of genuine contemplative silence or mental vacuity or absorbing distraction, step firmly on it by saying the Mantra. Do not waste time or effort analysing it. Just go back to your Mantra.

Distractions
Attention or attraction is to be pulled in a definite direction. Distraction is to be pulled in other directions. When we try to meditate, to be still, we find ourselves being distracted. Our desires, our regrets for the past, our plans for the future, all pull us in different directions. The Ego is ever seeking to be lord of the manor. The mantra tradition gives us a remedy for this. When we get distracted we know exactly what to do to come into traction, just begin again to say the mantra.

Ultimately 'noise' is always internal and we must learn to be silent even in the market place. There was a monk in the city monastery who envied those in the country one. There, he said, it would be easy to meditate as there were no distractions. He got permission to go to the country monastery. He settled down to meditate and noticed that the fragrance of the Joss sticks, as they burnt before the altar, was much finer than in the city. It would be possible, he thought, to get some from the country to use in the town. In fact, it may be possible to arrange with the person who delivered the groceries to bring them on a regular basis and they could even sell them in town and make a little income for the monastery... The time for meditation was over. He realised that the external place has little to do with silence.

Posture
All traditions agree that good posture helps meditation. One essential rule is that the spine should be straight. As lying down is often associated with sleep it is not normally a wise posture but Fr Bede Griffiths recommends it, especially for the elderly. If the head, the chest and the abdomen are imagined to be three boxes, a good posture is to have these sitting nicely one on the other. I prefer to use a prayer stool as it helps to make those boxes sit properly on one another. A definite prayer posture can be a prayer in itself and by association dispose one for meditation. Any posture in which the knees are higher than the thighs is not helpful.

Time
Two periods of meditation, one in the first half of the day and the other in the second, are essential to the teaching and tradition about meditation. The minimum time is twenty minutes and the optimum time is thirty. For children it is recommended that they meditate one minute for each year of their age. So a six year old would meditate for six minutes.

It is important to meditate for the planned time, to not abandon meditation early if things are going badly or extend the period if you feel ecstatic. As mentioned already we meditate to be faithful not to be successful. To indulge a 'good' meditation, or abandon a distracted one, would be to follow 'success' and the Ego.

According to one's circumstances in life one has to find the best time to meditate. Some can meditate when they wake up during the night, others in the early morning before the bustle of the day begins.

Meditators have found that family members, friends and business associates can be quite supportive and accommodating if one has the courage to say that one wants some time apart to meditate.

Most traditions teach that it is not good to meditate soon after a meal. Personally, I have not found that this makes any significant difference.

Breathing

Good breathing helps meditation, and meditation helps toward good breathing. Some traditions just focus on the breathing and this has for them the same effect as saying the mantra. Most people find that they will tend to say the mantra in time with their breathing, for example, MA-RA with the in-breath and NA-THA with the out-breath.

Smooth breathing in and out is important for health. When we are fearful we tend to hold the breath. This happens dramatically in the person who is prone to asthma. Meditation can help people with this condition. It also helps to reduce blood pressure although these effects are not the purpose of meditation. As a scuba diver I found that my air consumption was reduced by about 25% after I began to meditate.

Expectations

In prayer directed to the 'God out there' we generally have expectations that this God will intervene in some way, and maybe even work a miracle, on our behalf. In prayer that is listening to the God who is present in the community, we are expecting God to give us wisdom, guidance, and courage as we work to empower people to bring about social transformation.

We meditate to be totally open without any expectations. By being in silence we are open to our own spirits and to the Spirit of God as they meet at our centres. This leads to a fullness of life that may lead to a life of dedicated community service.

Expectations already limit openness. My first teacher in meditation used to say that if you have expectations in prayer and they are not fulfilled you have got what you deserve!

A student went to his master and asked for the mantra that would give him enlightenment. The master gave it to him. He used it for fifteen years without reaching enlightenment and went back to complain to his master. The master asked him what mantra he had given. On hearing it, he said, 'Oh sorry, that was the wrong mantra to bring you to enlightenment. But it is too late now to change. Just continue with the one I gave you.' He continued, and in a very short time enlightenment came!

Meditation Groups
Most people find that they are helped a lot by belonging to a prayer group. Some find that they meditate better by just being with others. I find that there are a number of other benefits also from being with a group. There is a sense of support and there can be a sharing of tapes, books and resources. More important is the sharing of the experience, especially the experience of 'failure'. We do not have to be a 'success' in meditation in the sense of getting rid of distraction. We just have to be faithful in coming back again and again to our mantra. Hearing the struggle of others convinces us that we are normal and should not give up just because we seem to be getting nowhere.

Most meditation groups begin with a tape or a talk, then meditate and finish with or without a discussion. In my groups we have three stages. I believe that these are very important for good theological reasons.

1. *Bible Sharing*. We usually start by sharing how the Sunday gospel has touched our lives. For most of us our God and religion is up there in heaven and in our heads. By sharing the scriptures and where they transect with our daily lives we are pulling God out of the sky and into the world around us.

2. *The Meditation Proper*. Here we BE with the Spirit who dwells within us.

3. *The Sharing*. Here we deal with the experiences and problems and misinformation that people have. Here also we get the encouragement to 'keep on keeping on', as John Main liked to say.

Is Meditation the same as Contemplation?
It is important that we know what we mean when we use the words meditation or contemplation as they have different meanings in different traditions.

For a long time in western spirituality, there were four stages of prayer; *lectio, meditatio, oratio* and *contemplatio*. *Lectio* meant to read something from scripture or a pious author. Then there was *meditatio*, a thinking or rumination on what was read. This led to *oratio* or a raising of the heart to God and possibly the making of acts or resolutions. Then there was *contemplatio*, a living in the presence of God, literally, under the same roof with God.

For the mystics, contemplation has meant a form of prayer where the senses are suspended, and meditation has meant prayer in which there was much reflection on a reading. In the Eastern traditions they had the opposite meanings.

St Ignatius and the Jesuit tradition use the word contemplation to mean entering imaginatively into a scripture scene.

In its root meaning meditation means to stand in the middle or 'to be at one's centre', contemplation means 'to live under the same roof with'. In John Main's teaching they would have these meanings and be almost interchangeable.

Is Meditation Christian?

By meditation we become more present to our own centres, we centre on the centre of Reality. When this happens we are implicitly or explicitly in the presence of God. Any healthy tradition of meditation does this.

For Christians, however, meditation is greatly enriched by the whole theology of the Trinity. We are told by St Paul that we do not know how to pray but that the Spirit prays within us ever crying, 'Abba, Father'. When we meditate we are present to the prayer of God, the community of the Trinity at prayer within us. There is a hint of this also in the request of Jesus to the disciples in the Garden of Gethsemane, 'You watch, while I pray'. This is the purest Christian vision; prayer is being in the presence of the Trinity at prayer. We are at the centre where the Trinity prays. We ourselves do nothing.

Is Meditation Dangerous?

In almost ten years of personal experience of meditation following the teachings of John Main and of working with meditation groups in different parts of the world I have never come across anybody that was hurt spiritually or psychologically by meditation. I believe that there is no danger in this form of meditation.

From history there are lingering fears about meditation. Protestants have told me that they were taught that if you meditate Satan will come in and take over your soul. Obviously, this is the product of contention between traditions where there may have been unhealthy practices involved.

From experience I can understand how these fears may have come about. As one meditates one becomes more honest. This means that repressed forces within may be released. Some of these are good some of them are bad. Anger, repressed hurt, may bubble up to the surface. This can be frightening and can be why some people are afraid to meditate. According to the tradition and vocabulary with which one has grown up these strong forces from within may be described as angels or demons. A meditation group needs a competent director to help understanding and discernment in these matters.

In the sixteenth century Quietism was condemned by Pope Innocent XI. According to the quietists salvation came entirely through God's action without our efforts. For them meditation or contemplation was the only kind of prayer. The fear of being branded a quietist has influenced the teaching of prayer for centuries. For example, St Alphonsus Liguori, a Doctor of Prayer, is very wary when he talks of contemplative prayer.

Mysticism, the direct unmediated communication between God and the individual, is a fact of Scripture, of history, and of the experience of many ordinary people. However, it is very often extremely difficult to discern what comes from God and what comes from a psychologically unstable mind or other influences like the use of drugs. Mystics tend to be vague, not fitting into a clear categories, and they often tend to have disturbing things to say to those who are content with the status quo. For these reasons the institutional Church has always been wary of mystics and mystical experience.

In our own times there are meditation teachers, and schools of teaching, whose motivation and practice merit critical scrutiny.

The gift of God is free but some 'teachers' demand big fees for the meditation word and technique that promise 'transcendence'.

Technique and Discipline.
For John Main the distinction between a technique and a discipline was very important. In many popular forms of meditation emphasis is put on techniques that will lead to peak experiences. These have no place in John Main's teaching. A technique is something to be mastered. To master it is to be successful and so to boost one's Ego. This can be spiritually counter-productive.

Discipline, on the other hand, is following a way. The discipline of meditation is external and internal. The external discipline is to meditate twice a day; the internal one is to keep coming back to your mantra no matter how often you become distracted.

We follow the way, not in order to be successful, but in order to be faithful. The measure of success, if there is such a measure, is our faithfulness in returning to the mantra, not our success in staying with it.

This is beautifully captured in the words I have already quoted from T.S. Elliot:
> There is only the fight to recover
> what has been lost
> and found and lost again and again;
> and now under conditions
> that seem unpropitious. But perhaps
> neither gain nor loss.
> For us, there is only the trying.
> The rest is not our business.

The Ego

There is an Indian story about a king who was given a totally dutiful servant. The servant would perform, indeed anticipate, all of his wishes, and when not given a task would be restlessly demanding things to do. At first the king was delighted to have such a wonderful servant who relieved him of many chores. After some time, however, he found that providing chores for the servant was becoming more exhausting than the chores themselves. Then he got an idea. He had the servant erect a forty foot pole in the garden and instructed him to climb up and down when there was nothing else to do. In this way the king got his freedom back.

The Ego is a little bit like that servant. It is part of us, an essential and useful part of us, but it can also be a demanding tyrant. When we say the mantra we are sending it up the pole and giving ourselves space from it. Perhaps the greatest thing that Christian meditation does for us is that it helps us to deal with the Ego.

I would like to define the Ego as the selfish Self in contrast to the true Self which is the selfless Self. (This may be different to Freud for whom the Id was the basic drives. The Superego was the 'do's' and 'dont's' to which we have been conditioned and the Ego was the referee between the two.) The Ego, as we use it, is following Jung's understanding of it. The true Self and the Ego are part of the light and shadow within us and through meditation we learn to hold them in balance. We learn to moderate the Ego which will bully us if not checked.

Our Egos are part of our upbringing. We have all been fixated, at least to some degree, at some childhood level of our development. Our past experiences give us Egos that are fearful and defensive because of the ways in which they have been indulged or deprived. This fearfulness hinders our ability to let go in transcendence; to see and live values for their own sakes; to love and seek justice without having a selfish motive. It smothers our true selves, our better selves, the selves we were called to be. The Ego is dualistic. There is me (Ego) and other objects. It is observing, fragmenting, comparing and measuring everything in terms of what is in it for me. The true Self on the other hand is whole, it seeks to integrate, to be one with the other, to have Agape, to love without strings attached. The Ego makes us love things and use people, the true Self challenges us to love people and use things.

Our Ego makes us desire to be great and if it is not transcended it will stop at nothing to achieve that greatness. The Ego makes us want to grab recognition, position, title, approval. It makes us slaves to the opinions and approval of others. The consumer society thrives on the Ego. It makes us fear not having the right car, or smell or gadget. The Ego loves shooting others down. It thinks, in some neurotic way, that it lifts itself by putting others down. It makes us jealous and competitive. It makes us feel threatened by the success of others.

The race horse 'O Leo' is a picture of the Ego. He was a very fast horse but he had one big fault. As he led down the final straight he would have to look back to see if any other horse was near. When he did this, invariably, he was passed on the other side.

The Ego is very stupid, very short sighted, it cannot see beyond its nose. It also has an insatiable appetite that is never satisfied. Feed it and it needs more feeding. Recently I heard the remark about someone, 'Feed his Ego enough and he will work for nothing'. The Ego does not know what is worth dying for. It is forever making mountains out of molehills, getting excited over non issues.

The Ego is a little bully. Most bullies are cowards at heart. When you stand up to them they back off. This is what we do when we meditate. We just stand our ground and the Ego is exposed and backs off.

As we meditate we first of all become more aware of the Ego. We become aware of when we are acting out of the Ego. You may notice for example that you are getting angry. Before this, anger was just a reaction. It happened to you. What was repressed just squirted out at an unexpected moment. For the one who has been meditating there is a moment of realisation. This anger is not because something is wrong but because my Ego has been challenged. With this new awareness you have a moment in which you can choose to give in to the anger or you can set it aside. You can be mastered or you can be master.

Or this may even happen unconsciously without your noticing it. One time a man was telling me that he had been meditating for a year but nothing had happened. The children contradicted him, 'When that lady backed into us this morning you just went out and discussed the damage with her. If that was last year we would have learned a whole lot of new words...!'

The Ego is not you. It is only a part of you. If you give it control, you hurt others, you do a lot of harm. If you just observe it you can grow. When you listen to the Ego it tells you who you are. It can be your friend. For example, I have learned a lot about myself by listening to the times and places that I have felt inclined to lie. Why did I feel like lying here? Why was I threatened? Was that a neurotic fear or was there a real reason to fear? As I face the fear more and more I get free from it. When we recognise the Ego as a part of our external personality we can distance ourselves from it and we can then be joyful free people. I like this story attributed to Aristotle Onasis.

A friend went to him and said 'My daughter, Martina, has Leukemia. I need $20,000 for treatment. Can you help?' Onasis wrote a check for $20,000. A few days later another friend told him that it was all a con job. Onasis simply replied 'I am so happy to hear that Martina is not sick.' I wonder if that would have been my response even after several years of meditation!

Meditation and Washing Socks

For many years now I have been hand washing my socks every few days. In those years I have learned a trick or two. At first I used to dip the socks in water and swirl them around for a while. Then I would hang them on a line or put them on a window sill. This process did not have much effect on either the socks or the water. Later I learned that it was advantageous to use soap. I would soap up my socks and scrub them between my fists and get soap up to my elbows. The water became dirty and I presume the socks became somewhat cleaner and I became a little more tolerable in company because of this process.

Then I discovered detergents: I just put the socks in a basin and added a little detergent. I would come back some time later and most of the work was done. There would be a coat of dirt at the bottom of the basin equaling presumably what had been dislodged from the socks. The thicker that coat of dirt the more satisfied I was. Then a few more rinsings and that was it. I seemed to be doing nothing and yet the job was being done.

What a wonderful paradigm of my journey in prayer. For a long time my prayer was dipping into the Lord. I tried to be with him but little washed off from him to me or me to him. He and I remained much the same. Bit by bit I discovered better ways of engaging him. And then I discovered John Main and the mantra and in some ways I was never the same again. I had found a way home. The mantra is the detergent and the dirt at the bottom of the basin is the flood of distractions that beset the meditator. The more the merrier. The more dirt there is, the more effective the detergent has been. The more distractions that emerge in meditation the more the meditation is being effective in dislodging Ego material. Distractions are always coming from our Egos, our pride and vanity, from our regrets about the past and our desires and daydreams about the future. The more they surface to consciousness and we let them go, the less of them are left behind. So to see distractions flying off, is like seeing the dirt at the bottom of the basin. They are a sign that the meditation is working.

The mantra, like the detergent, is very necessary. Just as I was for a long time ineffectively washing my socks so too I was for a long time praying ineffectively. I was saying prayers but not engaging with God or maybe not engaging or disengaging with the opposite pole, the Ego. We cannot grasp God. But by being filled with Ego we make it impossible for God to be in the space that he could occupy. If we can dislodge the Ego from that place God will be there as it were by default. He will have a true home at our centres, 'as it was in the beginning.'

There can be a kind of prayer, even a kind of silence, where we are fearfully sitting on the hatches, trying to keep everything down, everything under control. When we say the mantra it is like saying 'ah' when a doctor is doing a throat examination. It opens the passage ways without our realising it. When we say the mantra there is a taking of attention off ourselves that allows the passage ways to the soul to open and what is repressed inside tends to flow out. When this happens we may try to close down again or try to grab each distraction as it comes out and grapple with it, or we can just let it go – good riddance – and feel the relief and joy of its exodus.

Once I was on an inter-island boat during a storm at sea. As the wind grew stronger the boat listed and seemed to be in danger of capsizing. The passengers and the crew tried to tie down the canvas side coverings to keep out the wind. But then the captain gave orders that the canvas was to be taken away altogether. At first the passengers objected. They were frightened. They would get wet and their goods would be destroyed. But when the canvas was taken away the wind blew straight through, the listing subsided, and the danger of capsizing was averted. We often cling to what can be destructive for us.

When we meditate then we should be joyful when distractions gush forth. We should not engage them or grapple with them - just observe them and let them go. There is nothing particularly discreditable about socks getting dirty. They pick up dust from outside and perspiration from inside. If there is nothing to pick up and they remain clean, well and good. If they get dirty we face the reality and wash them. So too when we come to meditation, if there are no distractions, well and good. If there are distractions be glad to let them go. Like the dirt in my socks the more that goes, the less that is left behind.

But no matter how clean my socks are now, I know that as I travel the dusty tropical roads they will accumulate dirt again in a short time. So too as we leave meditation the forces of the Ego reconvene. Just as the washing of socks is a twice daily process so too the washing off of the Ego manifestations is a twice daily necessity.

A Wider Vision of Prayer

Is Meditation Prayer?

Many people are frightened of meditation because they doubt if it is really prayer. I would distinguish three ways of praying:
Prayer that uses words, either vocal or mental;
prayer by listening, to scripture or to life;
and the prayer of simply being.

Traditionally we have been taught that prayer is talking to God. This leaves us with a feeling that if we are not talking we are not praying. But if we believe that God is the centre of Reality then we are making a great act of praise and worship by simply and attentively being in his presence. In fact, words and images can even be an obstacle to being in this presence. When we use words we are limited by them because they can only express a fraction of reality. We do not imagine what is present, we are just present to it. Images indicate distance from. While images can help in certain types of prayer, in meditation images are set aside. We just try to be, to be at our own centres, to be where God is. By saying the mantra we take the focus of attention off ourselves, we move our selfish selves away from the centre of the stage. Then, as it were by default, we give God his rightful place. This is prayer.

Recently, a doctor told me that one of the most difficult things to do medically is to do nothing. When faced with a patient one is faced with one's own need to have an answer or do something, the patient's expectation that you do something, the pressure of drug companies to use their products, the fear of a negligence lawsuit if somebody claims later that you should have done something. Yet often the best thing is to let nature take its course. It is becoming harder and harder to do this. I think it's the same in prayer. We are pressured from outside and inside to do or say something while it may be best to just be still and silent.

Prayer Beyond Talking

Recently, I listened to an audio tape on the Rosary brought out by an Australian Redemptorist. In an otherwise responsible presentation he dropped this clangor 'the mantra is just a modern fad'. In my earlier days of commitment to the mantra this might have made me angry. I now had a good chuckle at it. However, it did bring home to me how easy it is to debunk ways of prayer with which we are not familiar. I am not sure that any way of praying is totally wrong, but I do know that if I meet someone who is totally convinced that his or her's is the only way to pray then I have met someone who is definitely wrong.

I believe the essential attitude to prayer is found in the first Book of Samuel, Chapter 3.

> The boy Samuel ministered to Yahweh under Eli's care. The word of Yahweh was rare in those days and visions infrequent.
>
> One night, Eli, who was by then half blind, was lying down in his room. The lamp of God was still lighted and Samuel was lying down in the house of Yahweh where the ark of God was located. Then Yahweh called, 'Samuel!Samuel!' Samuel answered, 'I am here!' and ran to Eli saying, 'I am here, for you called me.' But Eli said, 'I did not call. Lie down again.' Samuel went back.
>
> Then Yahweh called again, 'Samuel!' and Samuel stood up and went to Eli saying, 'I am here, for you called me.' But Eli said, 'I did not call you, my son. Go back and lie down.'
>
> Now Samuel did not yet know Yahweh and the word of Yahweh had not yet been revealed to him. But Yahweh

called Samuel for the third time. Samuel stood up and went to Eli saying, 'I am here, for you called me.' Then Eli realised that Yahweh was calling the boy and he said to Samuel. 'Go. Lie down, and if he calls you again, say "Speak, Yahweh, for your servant hears."' So Samuel went back and laid down.

Then Yahweh came and stood before Samuel, calling as he did before, 'Samuel! Samuel!' And Samuel answered 'Speak, Yahweh, for your servant hears'.

The attitude of hearing, listening, being attentive, waiting on the Lord is the key to prayer. We can and should listen to the God whose kingdom is outside us, around us and within us.

While none of these divisions are watertight or mutually exclusive we address the God outside us, the transcendent God, the Creator God in words. They are the words of liturgy, worship, chant, music. They can also be the words of architecture, gesture, and posture. They express a reverence for and a dependence on God. This is the form of prayer that we mostly encounter in our liturgies, popular devotions, and in charismatic types of prayer. Listening to words can be a very wonderful kind of prayer.

Listening can often be difficult. The words used are not our words but those of another. In the past the words were often in Latin, a language most did not understand. So, sometimes we do not know what we are saying. Yet in prayer we should mean what we say and say what we mean. Very often prayer said with others goes too quickly for us to relish, to hear in our hearts, the beauty and richness of the words.

There is a great danger of this kind of prayer just becoming a formula. What happens in prayer can be compared to John and Mary who were madly in love. One day John, who was studying literature, came across some beautiful passages from Shakespeare's play *Romeo and Juliet* and copied them into his letter to Mary. She was absolutely thrilled as the words expressed exactly how they felt for one another. However, some time later she found he was doing nothing else except quoting from the play. She shredded his letters and warned that she might shred him too if that was all he sent! What once had warmth and freshness became a formula and expressed nothing. This can happen too in liturgy. What was once very rich and meaningful, can become boring and meaningless when endlessly repeated.

The traditional mental prayer or discursive prayer was more engaging. One was advised to read a passage from Scripture or a spiritual book and to reflect on it and apply it to one's life. One would then make some kind of resolutions. Many people became very holy through this kind of prayer. Today we would see it as very rational. The mind is in control, the heart is scarcely allowed to speak. Resolutions are less in favour now also. If one makes a resolution only two things can happen; it is kept or it is broken. If it is kept one has succeeded and there can be a spiritual achievement in this that is Ego satisfying. If one fails one is depressed, so one looses either way.

Today, examination of consciousness is much more in favour. In this form of prayer one tries to be aware or conscious of how he or she has heard, and has responded or failed to respond to, God's presence in his or her day.

The Roman Catholic tradition through history tended to get fixated on prayer using words. Until recently there was little place for any other kind of prayer.

Today we are more open to hearing God speaking in and through the world. God does this mainly in two ways; through nature and through history. We can listen to him in nature. The beauty of the sea, the sky, the land, the mountains, the flowers, the plants, the animals. We can hear him also in the pain of the devastation and pollution of the environment which is one of the major sins of our generation.

We can hear it also in the relationship of God with human beings. We hear this first of all in the written word found in the Bible. We also find it in life, the signs of the times, the movements and aspirations of people. This kind of prayer can be done in private, on a retreat, or in a neighbourhood, family or community gathering.

Some people can get in touch with deeper reality through pottery making, painting, or the active use of the imagination. The *Spiritual Exercises* of St Ignatius, in their original and renewed form, are the classic examples of this kind of prayer. St Ignatius wanted to 'find God in all things.' However, for many centuries the *Exercises* too became limited to their intellectual aspects.

The prayer of the Basic Christian or Ecclesial Communities in so many parts of the world today is a great example of listening to God speaking in history today. As the poor reflect on their situation in the light of the Scriptures they have read, God speaks his word with a new directness and relevance. This prayer leads to action. Having reflected on the situation in the light of the Word of God the prayer moves from the heart to the hands and feet.

The last kind of hearing is to listen to Being itself. To be before God without an agenda or a plan or a regret. In this kind of prayer there is a total letting go and opening up to God in silent meditation. This is mysticism in the sense that there is direct unmediated being in relationship with God. It rejects the contemporary fascination with 'peak experiences,' which is absolutely foreign to the wholly un-self-seeking attitude of the authentic mystic. The meditator shows not a magical trust in techniques, but a poverty, nakedness, and detachment from all that is superfluous.

In this book I emphasise this kind of prayer. I do not imply that it is the only kind of prayer. I simply share my experience that the more one meditates the more one becomes sensitive and consciously better able to listen to God's present word in people, events, scripture and liturgy.

Prayer Beyond Wanting Anything

People who criticise Christian meditation, but who have not themselves practised it, talk about it as if it were some new fangled idea, or as if it were a set of techniques for getting into peak experiences or into altered states of consciousness. When I hear that, I'm reminded of a story about a little boy fishing on a Sunday morning in the North of Ireland where Protestants are very strict on the observance of the Sabbath day. This Protestant minister saw the boy and said very reprovingly: 'Imagine catching fish on a Sunday!' The little boy looked back at him from the corner of his eye and said: 'Who said anything about catching fish?'

Christian meditation is a very earthy, tedious, unspectacular happening. It is just simply trying to say the mantra for twenty to thirty minutes twice a day, and that's about all there is to it. There may be some who have special experiences or find that their blood pressure has decreased as a result of meditation. If that's the gift they get, fine. But it is the journey, the process, that is important. It is not important to have achieved an end. It is im-portant to be where we are now and to be going in the right direction. When we meditate, we set out to say the mantra well. Actually, we might not be able to say the mantra well but we will notice that there are side benefits that come to us. If we go fishing we might not catch fish but we could have a lovely relaxing day with our friends.

When we start meditating, most of us begin for quite selfish resons. We find ourselves meditating because we hope it will help us to be calmer, to be more patient; we may start out of curiosity or because of the enthusiasm of a friend. As we meditate, the process itself purifies our motivations and our reasons for meditating. Ultimately, we meditate to become more God-like, to be more in union with the Trinity, to be able to live in perfect communion and to be able to live in perfect love.

Meditation makes us more loving people. It makes us people who can be and live at our own centres. It makes us people who respond to others, not because of what we can get out of them or out of a compulsive sense of obligation.

One of the big obstacles to loving is anger and irritation within us. We tend to get angry and irritated with others who don't conform to our standards, who don't do things the way we do, or who come along just at the wrong time and embarrass us or put us on the spot. When we get angry, we feel guilty and we're fearful of our own anger. So, anger often leads to fear. When we're meditating, we try to stay with our prayer word, our mantra, and whenever we fail and get distracted, we keep coming back. If we come back with anger, too bad. We have to learn to be patient in our half hour of prayer and to be silent. When we lose the silence, when we lose the word, we come back to it with patience and persistence, quietly and calmly, not getting angry, not blowing our top but just quietly coming back. We will then find that in the rest of life we will also be able to quietly come back and face difficult situations. When we meditate we find that our passions, our Egos, do not dominate us as they did before. We can seriously and quietly look at reality and respond to it more wisely, without violent feelings or thoughts.

Then meditation leads us to healing. We have a natural instinct to be always seeking balance. The negative feelings within us have an energy in them that wants to go back to balance, to go back to the positive. It is like when we see a crooked picture on the wall there is an instinct to straighten it. But what we often do is try to hide our negative feelings by putting a nice feeling over them or by thinking about something else. Often we spiritualise the problem, we offer it up to Jesus. This helps us to cope with the problem but it cuts off the process by which balance can be restored and healing can take place. Whereas, if we could just stay with the negative feelings, the healing process will take place. If we're in a bad mood, and we just stay with that and allow ourselves to be depressed or to be angry without doing anything about it, it will change itself to a better mood. But very often we drink, or take medicine, or try to jolly ourselves up with trivial stimulations. Then the natural healing never gets a chance to take place.

When we meditate in silence, in stillness, without wanting anything, the instinct to heal is given a chance. An up-righting can take place which cannot happen when we are denying our problems, racing around, thinking compulsively or talking incessantly.

Prayer Beyond Insecurity

One of the texts that was central to the teaching of John Main on Christian meditation was, 'I have come that you may have life and have it to the full'. Prayer should be a path to fullness of life, a fullness of joy within that pours itself out in love and compassion for those around us. Yet my experience has been that so many people live in fear, even people who pray a lot. For many their praying is a sign of their fear because they pray out of a sense of obligation. I experienced many people who spent hours at their prayers and who were themselves very unhappy people and brought tension and unhappiness to all around them. I was aware too of my own fears and unfreedoms that continued in spite of my prayers.

This led me to ask the question. 'What is the root of sin, of unhappiness, of alienation, of evil in our world?' I had often heard that it was sin or greed, or lust or the desire for power. Now, I think that insecurity is the root which drives people into grasping, showing off, lies, domination of others, anger, flying off the handle, panic, compulsive behaviour, drugs, drink and the misuse of sex. Many people become addicted to their own insecurity and suffering and project it, and the causes of it, onto the people and the world around them.

The Lord called us to fullness of life, to joy. The world in which we live continues to give us false recipes for happiness. It teaches us through our families, society and the media what will make us happy. We will be happy if we are healthy, powerful, rich, have what others do not have. It teaches that sex and money will give us happiness. It fills us with desires for things that we must be attached to, because if we do not have them we are going to be unhappy. The trouble with these recipes for happiness is that they are also recipes for unhappiness because they are saying that if you do not have, and identify yourself with, what they prescribe you will not be happy. If you are convinced that you will not be happy unless your partner is a millionaire you are going to be very unhappy if he or she has only 999,999. If you are convinced that you cannot be happy if you are not healthy then your life is in a shambles if you have a chronic illness.

Another problem about these recipes is that they put happiness outside ourselves. Your happiness, and also your unhappiness, is determined by having some object outside yourself which will make you look better before others. So the external object and others are what determine your happiness. But happiness is ultimately within oneself.

Another part of our conditioning is that there is a big God up there. If we do not play along with this God we are in trouble and will be punished like naughty children. If we play along with him he will give us what we want to be happy. So our relationship with God is conditioned by fear and greed. We fulfil our obligations to him to keep him off our backs and also so that we can wangle out of him what we want. Again our source of happiness is outside us and again this is a recipe for disastrous suffering. If we are full of desires and fears when we come to prayer and are using God to fulfil those desires and avert those fears we are destined for frustration and suffering, rather than growth and joy.

The recipe offered by Jesus was quite different. In the beatitudes he said, 'Blessed are the poor in spirit for theirs is the kingdom of God.' One translation says, 'Blessed are those who know their need for God'. Poverty is the secret of blessedness, of happiness. In another place Jesus said that if we are to follow him we must leave everything behind – yes! everything. Jesus also tells us that the kingdom of God is within us. His life, his joy, can only be found within our own selves!

It is the genius of John Main that he rediscovered a way for prayer in the Christian tradition that does what Jesus told us to do. By praying the mantra for twenty to thirty minutes twice daily we are leaving self behind. We are also leaving our desires and plans and attachments behind. We are no longer striving to fulfil our wants and being disappointed when they are not fulfilled. We are no longer slaves of people and events and objects or even of a God outside of us. Now we are transcending our attachments, desires and wants. We are just being with God no longer untrusting, insecurely telling him what he should be doing to make us happy, but knowing that it is enough to be with him and to be clay in his hands.

For me, the discovery of the mantra and the teachings of John Main and of Laurence Freeman was the discovery of a way to freedom, a way to joy, a way to transcendence of enslavement to the Ego and to what others think and impose. It was a road to integration of self, to the recognition of the passions that are within and of sorting them out so that they could be constructive rather than destructive in my life. The prayer of the mantra is a prayer that leads from fear into freedom, from insecurity to un-attachment and from fear of a transcendent God to joy in an immanent one.

Prayer Beyond the Fear of Prayer

When Mary goes to pray she has a list of prayers to say and of people and things to pray for. John knows that he should pray but he is always too busy. When Thelma tries to pray she invariably falls asleep and when Tony tries, he gets a backache. Tom finds his mind teeming with ideas when the time for prayer comes.

What do all these have in common? Of course, we cannot judge any person or say absolutely what is true for each one, but it is very likely that all of the above are afraid of prayer. Prayer is a call to be absolutely naked and honest before God, to be totally open, to be 'The handmaid of the Lord'. It is a call to say 'Your Kingdom come, your will be done on earth' and 'Not my will but thine be done'. If we are not ready to hear this call we will protect ourselves in different ways.

Mary tried to do it by control. She knew what she had to do and she knew what she wanted from God and so she was able to eliminate any possibility of hearing what God wanted. She was probably praying because it was an obligation that she had to fulfil and there was very little about relationship in it from the start. John had probably a better sense of what prayer was, of how it would challenge him. But, he was like the whiskey priest in 'The Power and the Glory' by Graham Greene, 'he wanted to pray but prayer is to act and he was not ready to act'. He was afraid of the challenges that prayer would make him face up to. Sleep is another way of avoiding self confrontation. When we do not want to hear what God is trying to say to us, we just fall asleep or we develop pains and aches.

When we want to avoid something we get a pain in the back or a headache, and if we listen to the pains they will tell us why we are avoiding prayer. Maybe we do not want to make a big decision.

Maybe we are in a relationship that is not healthy, or there is someone we will not forgive or we do not want our business practices to be questioned. Whatever it is, there is an area in our lives with a big 'No Entry' sign over it and we do not want the Lord poking around there. The reality then is that we are afraid to pray.

One thing is very clear - if we are afraid we will not listen, either to ourselves, to others or to the Lord. The flip side of that statement is also true - if we listen we will not be afraid!

In my own experience, and in the experience of those that I have directed, I have come to know two ways of prayer that can take one beyond fear. One is by using the imagination and the other is by not using it. One comes from the tradition of using words and images exemplified by St Teresa of Avila and St Ignatius of Loyola. The other comes from the tradition of letting go of all things, exemplified by St John of the Cross and John Main, OSB. One is in the tradition associated with the West and the other is in the tradition associated with the East. One is from the left side of the brain, the other from the right brain side. Again, I have found in my personal and vicarious experience that there is no contradiction between the two ways. They are complementary to each other, as the two sides of the brain are complementary to each other. They can be used at different periods of one's life or they can be used at different periods of prayer even in the same day. Often the more active imaginative type of prayer will lead nat-urally, as time goes on, to the more passive, still type.

Spiritual direction can be great fun. If you are a person who appreciates a good story, you will hear some of the best in this process, particularly if you are using the Ignatian method of contemplation. According to this method, the retreatant is advised to enter into a Scripture scene. The retreatant visualises the time and place of the scene. It can be in an ancient or a modern setting. He or she may take a role in what is going on, hear everything that is going on and even ask questions of the participants.

'I was visiting my friend Mary,' said Cathy 'we had been at High School together and had been very close friends. Whenever I had a problem I told her and she likewise me. Well, as we were chatting, this big handsome angel appeared and said to me, 'Could you excuse us a moment but I have a message from God for Mary.' Well, I went out fuming. Why Mary? Why not me? This was Cathy's story. As she looked back over the story she got in touch with a lot that she was hiding in herself, her jealousy, her ambition, her poor self image, that made her want to dominate others.

Rex was a very well-behaved seminarian and he was asked to meditate on the journey of the Magi. He found himself in the story as the trusted servant of Herod. He heard the whole investigation of the Magi and then asked to be able to follow them to find and eliminate the Child. As I said, Rex was a very well-behaved seminarian, yet reflecting on this story he realised that he was serving a Herod all his life. He had been beaten by his father when he was a child and had subconsciously decided that the way to get on was to be servile towards authority. So he was always good and obedient towards authority at home and later in the seminary. He had even internalised this so much that he believed that all nonconformists should be punished severely or eliminated. Rex was terrified of facing the repressed anger and latent violence within himself, but through his imaginative prayer and his sitting still afterwards, he was able to find, accept, and deal with his true self.

Imaginative prayer is a great means of facing reality and therefore, of facing God, the centre of reality. Some spiritual directors help their directees by dream analysis, by pottery making, painting. The key thing is that reality is faced, that it is not avoided or covered over.

Meditation is another way of praying that helps us to face and overcome the fear of prayer. In meditation there is no welcoming of the products of the imagination. They are seen as the product of the Ego which must not be allowed to control the person. In meditation we do not try to think of anything, to imagine anything, holy or otherwise. We just be. We neither fight nor take flight. We just stand our ground. There is a marvellous scene in the film

'Gandhi' where the people face up to the British troops. They accept being beaten by truncheons but they will not fight back. This non-violent approach led eventually to the demoralisation of the imperial forces in India. This is what we do when we meditate. When we meditate we just be, we hold on like a survivor hanging on to a lifebelt during a storm and eventually the storm passes and we are left stronger people.

Many people are afraid of meditation because of what it may bring up. It is impossible to be silent and to continue to be dishonest or angry or unforgiving within oneself. Either you give up Meditation or you let go of these emotions. Because so many people have actually been reared on a diet of hurt and anger they are not truly living. They are consciously or unconsciously blaming some person or event in the past for their present state. The hurt may or may not be true, but to continue to blame someone else for it now is to say 'I can do nothing to better my situation because of what ... they... did to me.' It is a dead end. Unfortunately, if one has been in this state for too long, it is the only place where one can be comfortable. We become dependent on these painful self justifications. The challenge to get out of this trap and live is frightening. In meditation the healthy self will tend to surface heralded by distractions and imaginings. As the meditator lets go of these and just continues to be, the inner space becomes less congested. The person will begin to loosen up inside and let go of the protective and compulsive behaviour that he or she was trapped in.

Prayer is relating with God, the God who called us to life and to have life to the full. It is a call to a God who is love, and love is what drives out fear. Unfortunately, prayer has often become a way of reinforcing fear instead of freedom. It has become a way of avoiding reality instead of facing it. We need the courage then, to pray in ways that help us to deal with our fears. I suggest contemplation in the Ignatian sense of entering imaginatively into a Scripture scene, or in the classical sense of being in silence under the same roof with God, as two forms of prayer that will bring us beyond the fear of prayer and beyond fear itself.

Prayer Beyond Aligning With Power

There were two things that people said about Sister Maria: she spent remarkably long hours in prayer and she was, to put it charitably, desperately hard to get on with. Generally one's prayer life, one's relating to God, is pretty consistent with the rest of one's life so we can make a guess at what she was doing when she prayed.

When she was not praying she was either fawning on those above her or tormenting those under her. She had a way of charming bishops and parish priests. She was so courteous, self demeaning, and theologically with it. She knew all the right answers and was able to convince those in authority that she could solve all their problems for them. She got herself put in charge of family life or liturgy or prison apostolate on parish level and sometimes on diocesan level. And then she went to work.

By the power (extricated by her or invested in her) by ecclesiastical authority she set out to 'free the people.' All would be well until someone expressed an idea. This was ignored if it was an idea contrary to hers, and the person was quickly put in place. After all she had the authority of the bishop behind her and what's more she had prayed over what she was doing and so had the backing of God almighty behind her. So who could tell her anything?

It would seem logical to deduce that for her prayer was aligning with power, the power of God. She was fawning on him to get his backing and then using that power to assert her own Ego.

This is not an uncommon form of prayer. It can be the prayer of the parish priest who convinces himself that, by fulfilling the obligatory prayer times and a little more, he is right with God and then he can take it out on everybody else. Parish Council presidents and leaders of church groups sometimes show the same tendency. Parents sometimes think that once they have said a few prayers they now have power to dictate to all.

Behind all of this is a fear mechanism, a seeking for security that is very far from what true prayer should be. I can describe it because I can identify with it and recall it in my own behaviour.

In my efforts to meditate over the years and in the constant education that I receive from the daily interchange with parishioners, I have seen that the first essential of prayer is to get away from power and to embrace poverty. Freedom is to learn that it is okay to be weak and powerless. It is fine for others to have other opinions. It is fine for them to have negative views of your projects and even of yourself. It is fine to be put under scrutiny. Criticism, if coming from the other person's anger or jealousy is their problem, otherwise it is valuable feedback to be received gratefully. So, why be upset about it?

This was the message Christ was constantly trying, and it seems failing, to get across to his apostles. After Peter's confession of faith he tries to tell them about the suffering involved in following him but they cannot hear it. Then there are constant squabbles about precedence, about who will be first in the Kingdom. They squabble at the Last Supper. They are even hoping, as he ascends into Heaven, that 'even now he will restore Israel.' They are hoping for a solution by power, but the way of Jesus is through poverty and conscious suffering. His attitude is entirely different and in my experience the best path to this attitude is the path of the Christian meditation.

The effort to say the mantra for 20 to 30 minutes each day is itself one of the greatest experiences of poverty. In ten years of meditating I doubt that I have ever been able to stay with the mantra for as long as two minutes. It is always a losing struggle. It is an experience of being a child, being powerless in the face of might. Yet, somehow God's strength emerges from this surrender to weakness.

Prayer Beyond Words and Theories

In my parish work, I have a lot to do with the sick. In the course of a few days I hear many reasons for sickness. One will tell me he is sick with an infection or that his resistance was weakened by germs. The 'hilot', the local healer, will say there's wind somewhere inside the patient's bones. Somebody else, who thinks that sickness is a punishment for sin, will ask what sin did he commit for which he is being punished. Another will say that he has this neighbour that hates him and that he has been a victim of sorcery, or that he might have offended a sacred tree. Others will tell me that they got sick because they were sleeping where water lines crossed under the ground. So, there is a limitless number of explanations for sickness.

Then, if we talk about good and bad, some people will say it is good to have communion in the hand, others will say it is bad. Some will say it is good that priests and nuns are dressed in special garb, others will say it is bad. So, every time you have opinions or points of view, you also have the beginning of controversy. Yet, the reality is only one. The theories can't all be true. Maybe they each have bits of the truth. So too when we talk about other things, what is good or bad? What a financier might see as very good, a social worker might see as very bad. What an engineer or an architect might see as very good, somebody else might find impractical or very bad.

When we're into words and theories, we're always in a sense chopping up reality and getting it wrong. The same thing happens when we talk about God. Anything we say about God is ultimately more wrong than right. For example, when we say God is our Father, the idea of fatherhood gives us some little pointer towards God but he's not a father (neither is he/she a mother) because he is not human. When we say that God is loving and almighty, these are human thoughts and they do not get anywhere near describing God.

There are so many different religions and so many different points of view on God. One will justify violence, the other will not, and so on. When we're into words and theories, we're always getting it wrong. But reality is one, being is one, God is one. There's something beyond the theories and beyond the words, there's Being, there's Reality pure and simple. When we pray without many words, without theories, without expectations, we're just Being with the Lord, who is the Centre of Reality.

I remember watching a touching scene on television. This boy had been born blind and he had just had an operation. There was a 95 percent chance that tomorrow, when they took off the bandages, he would be able to see for the first time. And he said to his girlfriend: 'I have impressions of you and I'm afraid that when I can see you, you might be completely different. I might be all wrong about how you look.' But she said: 'Darling, it doesn't matter if the impressions are all wrong, our love is still the same.' There's something deeper than impressions and that is love, agape. And I think this is exactly what we're aiming at in meditation; to be beyond impressions, to be beyond theories, to be beyond words.

Words can often be a way of expressing fear and blocking and limiting reality. If there are 'No Entry' areas in our lives, areas into which we do not want the Lord to probe, we can keep him out by using a lot of words in prayer. In pure prayer, in meditation, we just let God be God.

Prayer Beyond Infringement of Space

The matter of space is very important internationally. When there is aggression on the space of one nation by another, it can make the whole world go to war. It is also very important for all of us. Neighbours can quarrel over space if one infringes on the property of another. Building a wall or dumping garbage into the property of another can cause deep hurts, deep pain between neighbours. Each of us personally needs space, we need physical space, the right distance between one another. When people sit down with one another they often adjust their chairs backwards or forwards to find just the right distance. Some people need more space than others.

We also need psychological space. We feel a need for others to respect our privacy and our time. If people demand our time when we are busy with something else or if they demand too much of our time, or if they are nosey about things that we consider private, we can feel that our psychological space has been infringed. Invading my space is invading my person and it can make me feel desecrated or violated.

Sometimes parents can manipulate or pressure their child to do what they want and this can leave deep anger because of the child's feeling that his integrity has been invaded. We all desire to be true to ourselves and when somebody does not give us the necessary space we get angry. Often we need space to mourn a loss and can be greatly harmed if we do not get it. So this matter of space is quite important.

I believe that in coping with this matter of space, meditation is very important. In a sense, what we are doing when we meditate is creating an open space within us. We are removing thoughts and images, we are just being by ourselves with the mantra. The Russian school of prayer talks about the Poustinia. It is a little house, a space, in which the meditator lives. When we meditate we enter our Poustinia, an open clear space within ourselves, and look at the rest of the world, from this sacred space, unperturbed. Gradually, the whole world becomes our Poustinia. When we have peace within ourselves we can have peace with the whole world and we feel that the world will not infringe on us, or, if it does, that we can deal with or adapt to the infringement.

There are three important points about meditation and space. First, when we meditate, we have a better capacity to realise why we are angry when our space is infringed. If the infringement of our space is because of our Ego, we become aware that our Ego is being hurt and this awareness enables us to deal with it better.

The second point is that we are helped to express our feelings constructively when our space has been infringed. How we express these feelings is very important. We should assert our rights and express our feelings but we should not be aggressive in doing this. Aggressiveness comes from the Ego, assertion comes from a confident self acceptance. If we are aggressive we will probably cause even greater trouble. When we're aware of how much our Ego is involved, we will have a greater capacity to be assertive rather than aggressive. To be assertive is not harmful. To be aggressive invites aggression from the other and does not really help things at all. Meditation gives us the strength to assert our rights without being aggressive and gives us the wisdom to know the difference between them.

The third importance of meditation with regards to space is that meditation gives us the capacity to let go and accept gracefully when people encroach on our space or do not respond to our requests for space. If they refuse to give us what we want, we can freely let go of the desire for it if the matter is not important. While watching television with another person we may feel that the volume should be lower but the other person feels it should be higher. Both are in a sense legitimate demands but we feel we are being imposed on by having it too loud while the other person feels he's being imposed on by having it lowered to please you. When we meditate we come to realise that this is really a trivial matter and there is no point in breaking relationships because of this. We can be quite happy, having asserted what we feel, to let go.

There are two stories of Dietrich Bonhoeffer which are particularly remarkable when told side by side. When Hitler set up his own church in Nazi Germany his friends and students tried to persuade Bonhoeffer to join the Reich Church. They told him that by joining he could continue to teach and preach. To this he said 'one act of obedience (to conscience) is better than a hundred sermons.' He died for this conviction.

On another occasion he was in a hall where the Nazi salute was given. His companion, Ebernard Bethge, was horrified to see him give the Nazi salute with apparent vigour. To which Bonhoeffer said, 'Put up your arm, this thing is not worth dying for.'

This matter of space, when to claim it and when to yield it, is very important for our happiness and for our fulfilment in life. Meditation helps us find the necessary wisdom.

Prayer Beyond Empty Ritual

A little boy had a pet turtle that he loved very much. One morning he was distraught when he found the little turtle stone cold. To distract him from his distress his father said, 'let us have a funeral for your turtle'.

So they worked hard for hours making a little coffin and a little coach to bring the turtle to its burial place. When they went to place the turtle in the coffin he could not be found. He had recovered and wandered down the garden. The little boy's first reaction was, 'Kill him, Kill him'.

This story illustrates well the value of ritual in dealing with transitions and also the danger that rituals can become ends in themselves and lose or even contradict their original purpose and meaning.

We live in a world of rituals. Most of us begin the day with toilet rituals. We have rituals to greet people and to begin our daily chores. Sometimes we attend inaugurations, graduations, and flag ceremonies. Rituals are employed in both the secular and the religious world. They are rites of passage, they give a sense of importance and mystery especially to moments of transition. We greet one another with words or an handshake. This action is a physical happening but also pointing to a mystery, the mystery of relationship and friendship.

At baptism the child is formally accepted as a member of the communion of Christians. At marriage the commitment of man and woman to each other in love is ritualised and formalised. The formality is a way of underlining that this is no trivial matter but something that must be taken very seriously. This is followed through with rituals of love, moments of tenderness, Valentine cards, 'Get well' cards, or anointings in time of sickness.

The rituals for 'Goodbye' are very important. We ritualise the separation from the ones we love when we go on a journey. We also ritualise the final goodbye, the goodbye of death itself. Different cultures have different ways of greeting one another after someone has died. Then there are the rituals of the wake, the funeral service and the cremation or burial.They are ways of saying;
1) This person is now dead, I can no longer deny it.
2) I have done the proper thing, all that I can possibly do about the matter at the moment,
3) I must restart life without this relationship.
This is the necessary process of bereavement, of accepting, letting go and restarting life. It is living out the beatitude, 'Blessed are those who mourn, they shall have joy'.

Unfortunately the rituals of separation are being eliminated today, especially in the so-called advanced nations. We miss them when they do not happen. I heard recently of the shock registered by a visitor to Europe. He visited the hospital with his associate whose relative had died. Then they walked away and that was it. The hospital would dispose of the cadaver. Psychiatric writing today tells of the terrible emotional problems that result from the failure to ritualise and cope with the changes in relationship that death brings about.

But it is not only the death of a loved one, or an unloved one, that we have to mourn. If we are to be healthy we must also deal with the many little deaths that happen within ourselves. I may have to deal with the death of realising that I am not as clever as I thought, that I cannot get the schooling I desire, that I failed an exam, that I was jilted in love or divorced, that I did commit a heinous crime. Karl Jung said that many are suffering now because they did not suffer in the past. If we did not avail of the rituals that help us to work through the deaths of the past, our systems are clogged up and we cannot deal with the deaths of the present.

Rituals are important in pointing to mystery, to transcendence, to something beyond the here and now. Tribal people celebrate the rivers, the forests, and the mountains; planting time and harvesting; birth and death. The rituals that we experience in childhood become part of the stories that create our life. Today many rituals have lost their meaning and we need to create new ones. We are in great danger of throwing out the baby with the bath water. Pictures, statues, altars in the home, processions, first communions, the use of holy water all point to the sacred. Some of these symbols may not be meaningful to the present generation, or may even smack of superstition, but their absence leaves a void, a lack of the sense of the sacred.

Unfortunately, ritual can become bereft of meaning:

> A learned monk went to lecture on prayer at an ancient monastery. He noticed that a cat was scratching at the lattice work and he asked two monks to take the cat and tie it up.
>
> Two hundred years later another monk went to lecture in the same monastery. Noticing two stalls empty he asked where the monks were. He was told that they had gone out looking for a cat to tie up before they begin their session.
>
> In the intervening years the monks had written many learned treatises on the advantages of tying up a cat before beginning to pray!

It can also become an end in itself:

> The Tzar of Russia went to attend Mass at a monastery that was famous for its wonderful choir. There was one monk there who was totally out of tune. The abbot put him under the strictest obedience not to open his mouth in song during the visit of the Tzar. However, when the choir was singing one of its most comp-licated pieces the raucous voice of the out-of-tune monk broke in on the performance. The abbot was furious, he later castigated the monk and put him on bread and water for one month as punishment for his disobedience.
>
> That night the Lord appeared to the abbot and said, 'Abbot, it is you who should be fasting on bread and water. Today, while all of you were performing out of pride, one monk sang out of devotion.'

I would say that one of the greatest temptations that I experience is the temptation to make my priesthood into a career, to become 'the person who performs rituals.' This is a temptation for me as an individual and for the whole Church as an institution. It is so easy to become a performer and to lose the in-spirit-ation. If ritual is not coming from the Spirit within it will be empty ritual. This is true in all areas of life. If one is not present in sincerity to a kiss it can become the kiss of Judas. There are many who swear allegiance to the flag as they start their work, and in their work cheat their country and their fellow citizens.

The practice of meditation brings us back to the source, to the Spirit who dwells within. It helps us to be present within ourselves and to be present to the holy acts that we do. It also puts us in constant vigilance over our Egos who much prefer to be served than to serve.

Prayer Beyond Compulsion

During a time when the the newspapers were reporting scandals daily, from presidents to bishops, from financial manipulation to blackmail, from adultery to child abuse, a colleague said to me, 'I just do not know how a person like that could do a thing like that!' I did not know whether to treat him with admiration, disbelief or pity! From the quirky feelings and tendencies that I sometime find within myself, I find myself identifying more with St Paul in Rm 7:18. 'I can want to do what is right, but I cannot do it.' I find myself praying like Philip Neri when he saw the convicts being paraded to execution 'there but for the grace of God go I.' There are inconsistencies and alienated areas within all of us and we do need God's grace to be able to handle them in a non-destructive way.

In today's world there is much alienation, perhaps, even more than ever in the past. We are rushing, rushing, doing so many things that we can lose our centres. When we lose our own centres, we do not lose the need for love. Often because we did not have good role models in our parents or significant adults in our lives, or because we ourselves are not willing to invest the time and effort, we may not be able or willing to form loving relationships. When people are in love they take a lot of notice of one another and give one another a lot of attention. Very often then instead of loving another we try to get attention. And instead of intimacy we try to get intensity. This is very noticeable in the world today. We try to get love and if we cannot get love we get attention. A child will throw something down and the mother comes along and spanks the child. At least it is getting attention. It is not a nobody. Adults can also seek to avoid being nobodies by having the latest car, dress or gadget or by competing for prestigious positions.

People can seek intensity in extraordinary ways. For example, kleptomaniacs can be quite well off people. A politician's wife who feels ignored may go into a store and steal things. She finds an intensity of sexual excitement in doing this and in the thrill of almost being caught. There are a host of ways in which people can find excitement, sexual or otherwise. This gives them intensity and this substitutes in some way for intimacy or a deeper relationship with a person. For this reason a man may seek a homosexual relationship rather than a heterosexual one. It can offer intensity without the need for intimacy and relationship of which he may be afraid.

These intensity making incidents can quickly become compulsive. The events or actions are used to change our mood, to get us out of a depression or a feeling of alienation. At the beginning they tend to work but after a while people still use them even if they don't work. An alcoholic starts off taking a drink to cheer him up, and then later on, even if the drink is no longer effective, still drinks. He is now addicted to it.

The more alienated we are, the more we seek fulfilment in these sort of things, rather than in intimacy, in true sharing and love for one another, in love for God and in true love for ourselves. The more we are out of touch with ourselves, with others and with God, the more we will need this kind of attention and intensity. Likewise the more we are in touch with ourselves, with others and with God, the less we will need these attention-seeking, intensity-making incidents in our lives.

This is where Christian meditation is very healing. As we meditate, we first become aware of areas of compulsion in our lives, even in simple things like smoking and coffee drinking. Then we can admit them and, as we begin to feel more at home within ourselves, gradually let go of them. As we meditate daily we create a vacuum in ourselves and the alienated parts tend to come together slowly. What is disconnected becomes reconnected. Gradually we become reconnected with ourselves, we become still within ourselves, we become at home within ourselves, we can love ourselves.

If we can love ourselves, we can love others. If we can be honest with ourselves, we can be honest with others, we can share with others, we can share with God and this is what intimacy is all about. The more that we have this genuine self-love, genuine intimacy, the less we will have the need for attention-seeking or compulsive activities.

Meditation can help a lot towards having healthy minds in healthy bodies. The milder addictions and dependencies that we have are constantly being cured through our meditation. Those who need professional assistance will find their therapy greatly helped by meditation.

Prayer Beyond Asking For Help

One of my parishioners had two married daughters. One Saturday he went to visit them. After some small talk he asked the first how things were. 'Fine,' she said, 'apart from the fact that my husband is a rice farmer and he can do nothing at the moment because we do not have enough rain. Could you ever pray for rain when you go to Mass tomorrow.'

In the afternoon he visited the other daughter. Again he asked how things were with her. 'Fine,' she answered, 'except for the fact that my husband is a salt farmer and he can do nothing at the moment because we have too much rain. Could you ever pray for dry weather when you go to Mass tomorrow.' He came and asked me what he should do!

His problem brings up an interesting question. What is prayer for? Is it to change God's will, to make him conform to our plans? I am certain that this is often the popular understanding of prayer but is it the correct one? Often our prayers could be translated 'My kingdom come. My will be done on earth' rather than the opposite. On a more sober look I think we will see that prayer is not to twist God's arm to get him to do our will. Rather it is to help us to find his will and then to surrender to it. By prayer we listen to his tune so that we can dance to it. This is clear in the life of our Lord himself. He listened to the Father in prayer before all the major events of his life. In the garden he prayed 'Would that this chalice would pass from me,' and then he added, 'But not my will but yours be done' and that's the way it worked out. It was the Father's will and not his that was fulfilled. He too had the experience of apparently unanswered prayer. Prayer is meant above all to open us and help us to accept reality because it is only in accepting and being with reality that we can be with God.

Let's look at the case of Joe. When I visited him in the hospital he almost jumped on me with his eyes popping out of his head.

'Don't I look fine, Father. You can see that I am not sick at all. Isn't that right, Father. Anyhow they have a cure for everything now. Isn't that right, Father.'

From my clinical pastoral education and my pastoral experience I could see that I had before me a very frightened man who was denying the reality that he was seriously sick. I could see too that he was not now ready to deal with that reality. He was only forty two years old, he had a wife and five children and had just been diagnosed as having a serious form of cancer. I saw my pastoral task as being with him, hearing his feelings, and knowing that there was no need for me to tell him anything. He would sooner or later talk about his fear and then go beyond it. I visited him almost daily and after two weeks he could calmly accept that he was seriously ill and was beginning to talk about his relationship with God from this new perspective. He was also able to talk about practical things like his will and checking up on his insurance.

Then one day I came to see him and he had regressed almost to day one. Everything was going to be alright, his sickness was gone. He was sure a miracle had happened. When he calmed down I found out that some group had come to 'pray over' him. They had told him that if he had faith he would be cured. That was all he needed. He went back into denying reality again. I felt rather annoyed about what had happened but decided not to let this interfere in my relationship with Joe.

Every now and then a prayer group, or a group with a relic of a holy person, goes to pray over the sick. Often they visit the sick and dying who are in no way dis-eased. The patient is often quite at ease even with facing death. It is the pray-ers who are dis-eased and are working out of some need of their own to be God. This can be particularly horrible when they lay the line on the parents of a sick child. 'If you have faith your child will be healed,' so then, if the child dies, the parents are left with the double burden, the loss of their child plus the fear that the child died because of their own lack of faith.

As a pastor, I am really concerned about this kind of prayer which is often coming more out of the needs of the pray-ers than out of the need of the person being prayed for. It can be destructive and cause a lot of unnecessary pain. It often hinders a true Christian attitude and acceptance of God's will.

If we reflect on it, 'prayer' is often associated with unreality or a kind of begging. We pray for a sort of magical intervention. We pray to pass the exam that we have not worked for. We tell someone that we will pray for them, and then probably forget it, when we mean that we feel hopeless and cannot deal with the reality before us.

I believe that an attitude of letting go in full trust to God's goodness and love, the attitude that one comes to learn and live through Christian meditation, is the way to face the various realities of life and death.

However, I am talking about reality and prayer. I would not be honest about reality and about the ways of our unpredictable God if I did not acknowledge that sometimes those 'prayed over' or those who have touched 'the glove', do make strange and remarkable recoveries. Miracles do happen. We are not facing reality either if we limit the ways in which God can work.

A traveller going to pray at a famous shrine stayed for the night with a couple. They would not accept any payment for putting him up, but asked him instead to pray for them at the shrine that they might have the gift of a child.

On his return he visited them and said, 'I am so sorry to have to tell you that the Deity said that it is not your fate to have children.'

Some years later he passed that way again and to his amazement there was a house full of children. He asked what happened. 'Another holy traveller passed by,' they said, 'and we asked him to pray for us also. We soon had children. When we asked about our fate we were told that the work of prayer is to change fate!'

Prayer Beyond Miracles

If people find out that I have come from the Philippines they say, 'Oh, that is the land of the faith healers.' They have heard of it as a place where many people, on whom doctors have given up, go looking for cures. I know it also as a place where many people believe in the power of sorcery, the power of the curse or the action that can bring evil on another. Ireland too has had its curers, its moving statues, and its piseogs. The sheer multiplicity of reported visions and apparitions in our time is quite astounding. Miracles, visions, paranormal happenings are reported from all over the globe and from different religious traditions. According to Dom Bede Griffith OSB, the great spiritual leader who has been in his Ashram in southern India since 1955, that continent abounds in people who can work miracles, can bilocate and levitate. They have extraordinary powers which can be used for good or for evil.

Explanations for such powers also abound. There is a faith healer in my parish who can diagnose sickness by touching a person and praying and then he will 'feel' where the sickness is. He says that this is a gift given to him by the Infant Jesus. I know a water diviner who can do exactly the same thing but sees no need for a supernatural explanation. Dom Bede Griffiths in a recent lecture in London (and in his book *A New Vision of Reality*) says that all of these paranormal happenings, whether miraculous or malicious, come out of the psyche and are not necessarily spiritual or transcendent. Some people have special power in their psyches which can be used for good or evil. Hitler and Stalin would be people who used such powers for evil. Healers may be very spiritual people and God may be using them to heal but what happens is still basically a psychic phenomenon.

It is clear that Christ had great power and could and did often heal people. I have heard one practitioner of pranic healing, who sometimes performs extraordinary and instant cures, explain some of the miracles of Jesus by saying that Jesus released the 'Prana' the healing power, in the person.

Biblical scholarship, while not discounting miracles, would be unanimous that the stories of extraordinary happenings did not lose in the telling as they were passed on in oral tradition. Rudolf Bultmann would even say that the miracles were invented or played up so as to make Jesus as good as the Greek wonder workers in the eyes of the early Greek audiences to which the Gospel was preached. In the Gospel accounts, we often find Jesus apprehensive about healing. He refuses to work miracles to show off his power (Mt.4:5-7, Lk. 23:6-12, Mk. 8:8-13, Mt. 12:38-42, Mk 15:13-32). He tells those healed not to tell any one else (Mk 7:33, 8:23, 9:25). He does not want people to be coming to him because he heals or feeds the body. He wants people to come to him out of faith and love. Faith is something that transcends signs. 'Blessed are those who have not seen and who have believed.'

I think that this whole area of miracles and visions is one of major pastoral importance. The question is not so much if these things happen, but how helpful is it for Christian maturing that a lot of spirituality is directed towards bringing about extraordinary intervention or inducing special powers. I think that the penchant for this kind of spirituality comes to a great extent from the tradition of preaching. The stories told and the letters read out in the context of novenas and other popular devotions generally emphasised some supernatural or almost supernatural happening. This is still the bias in the teaching of many popular directors or directresses of clergy retreats. Prayer is explicitly or implicitly taught as a way of getting around or behind God's ordinary way of acting. It is often taught as a way of circumventing reality.

The remarkable thing about Jesus was not that he worked miracles but his 'Abba relationship', that he was in such total relationship with the Father. He was faithful to the Father and sought to do his will not only when things were going well but also when they were going badly. He was ever trying to find truth and reality and willing to face up to it. He was challenged to show his power by coming down from the cross. He always refused to prove himself by signs. His will was to do the will of the Father and in doing this he came to the peace and joy of the Resurrection.

I think that the idea of a journey or an evolving development is important in all areas of life whether physical, psychological or spiritual. The world itself is seen as evolving from matter to life, from life to rational life, from the rational to the spiritual and transcendent. The Bible begins with the vision of harmony in paradise and ends with St John's vision of a new heaven and a new earth, and the word Maranatha.

Maturity and balance, either emotional or spiritual, is not so much where one is at, as the direction in which one is going and the degree of integrated wholeness that there is between and among the stages. The child starts off body-centred and totally selfish. Physical pleasure and the avoidance of pain are of priority. Later it learns that pleasure and pain also come from other people. It learns to socialise. Later again it learns about the pleasure of the transcendent – of music, art and of values like honesty, justice and chastity. The physical and the social will still be valid sources of pleasure. A hot bath, a good meal, sex, an evening with friends are all wonderful, provided that they are not in conflict with transcendent values like truth, goodness, justice or chastity. Immaturity is to get stuck at a particular level and, to lose the freedom to respond to and move on to the transcendent.

So too there should be a direction and an integration in spiritual growth. When people start to pray they usually do so for selfish motives. They may pray because they were told it was something that they should do and they are afraid 'something might happen' if they do not. They may pray because they want some thing, even something like to be holy, to be calm, to be chaste. St John of the Cross warns that spiritual riches can be as enslaving as material ones. The main motivation is the fear of punishment or the gratification of needs. I am amazed at the number of otherwise mature people who have never gone beyond this approach to prayer.

Their prayer is to a God 'out there' whom they want to 'fix things' for them. Being a little fearful of 'him', they also try to get around him through their devotions to Our Lady or to a favourite saint. Prayer that aims only at the fulfilment of needs is a spiritual cul-de-sac, a dead end for the soul.

The Lord's Prayer, the model of all prayer, gives us a very different perspective. It begins by setting up priorities of reverence. The first priority is reverence for God who is our Father/Mother, whose Kingdom is to come and whose will is to be done before all else. After that we reverence ourselves and look to our daily needs, our daily bread. Then we look with reverence to our social relationships and the unending need to be letting go of hurt if we are to grow in these relationships. Our actual prayer often goes very differently. It would be more like 'My Father in heaven, holy be my name, my kingdom come, my will be done on earth... Give me all that I can get.... and change all those people who make life difficult for me.'

The pastoral challenge is for pastoral agents themselves to become more free in their own prayer and to lead others to similar freedom. Prayer is not just to fulfil desires but to help us to the peace and freedom of transcending them.

Let me illustrate this point with a story. This man was given three wishes and he said, 'The wife is breaking my heart. I wish I were rid of her and then I would be a happy man,' and presto! his wife was gone. A week later he was broken-hearted, he had not realised how much he had loved his wife in spite of superficial differences, 'Oh, if only I had my wife back I would be a happy man!' Presto! she was back. Now he realised that he had only one wish left and that he would have to be more careful about it. He asked advice. He was told by one to ask for money because money can buy anything and everything. But another said, 'What good is money if you do not have good health, the thing to ask for is health.' Yet another said, 'What good is health if you are to die anyhow!' In his confusion he went to the enlightened one and got this advice, 'Ask that you be contented whatever you get.' This is the freedom to which prayer should be bringing the Christian.

There is a place for petition in Christian prayer. The Christian who knows that he or she is loved by God, will present his or her petitions to the Divine in full confidence of a loving and caring response. The more confidence there is in this response the more every prayer will be answered. When what is asked is first of all God's will, what is asked will always be received, what is sought will always be found.

People have to face most desperate problems in their lives. They have been taught to turn to prayer as a way of achieving what they want to solve their problems. But what they ask for may not be what God wills. What seems right in their wisdom may not be right in his wisdom. True prayer is not for having our desires fulfilled but to enable us to transcend them. It is not to give us our wants but to enable us to transcend wanting. Prayer is not to make God conform to our wills but to free us to conform to his. Prayer is an act of being, of worship, before our loving God.

But we must have great reverence and respect for where people are at. We must reverence the faith and devotion of our people. If we hear of extraordinary happenings we must hear them through with reverence and discernment. God does allow marvellous and mysterious happenings. They may be miraculous or malicious. We need not get too excited about them because they are peripheral to the core message. We do not need to unpreach or trivialise or condemn people's beliefs. We need only offer them a better menu and when they find what is better they will let go of what is not so good. When one is aiming at transcendence one stumbles on peace and joy. These are found not in the fulfilment of desires but in the transcendence of wanting. That is why a form of prayer, like Christian meditation, that helps one to leave self behind, to transcend the Ego and wanting seems to be a more desirable starting point.

I was asked recently if Christian meditation was for healing. I think it is for even more than that. Real healing is to go beyond the desire for healing, just as real wealth is to be beyond the desire for wealth.

Prayer Beyond Frustration

Frustration is the feeling that we experience when our expectations are not fulfilled. Frustration is anger because desires have not been satisfied. It is fear of what will happen to us when the recipe for happiness that we have set ourselves does not materialise.

We can experience frustration with ourselves. You may be frustrated because you wanted to be a beauty queen or a football star. You may be frustrated because you are not as good or as competent or as well treated as some people around you seem to be. Envy and jealousy often lead to frustration. Or you may be frustrated because of your expectations of others, a spouse, a child, a religious superior. When they do not behave according to your plan for them you are frustrated. Stress arises when there is a significant difference between what we expect and what actually happens.

We can be frustrated also with God. Almost everybody knows what God should be doing and they spend their prayer time telling him. But he has his own plans - and dare I say, possibly wiser plans. I once heard an Irish friend describe another by saying 'He is as contrary as the will of God!'. God's plans are rather unpredictable, very contrary, and if we place our happiness in changing them, we are due for frustration.

Jesus told us again and again that following him would not be easy. After Peter was put in charge of the Apostles Jesus told him that they would go up to Jerusalem where he would suffer and be rejected. Peter remonstrated with him, considering this preposterous for the Messiah. But Jesus said to him, 'Get behind me Satan, your thinking is not God's thinking.'

Jesus said, 'Come to me all you who labour and are over burdened and I will refresh you, I will make your burden light.' – but it would still be a burden.

He who was God became human, yet he had to face every kind of frustration and eventually death. That pretty well shreds the expectation that we should be trouble free and worry free if we follow Jesus. Yet, we hold on to it and in holding on to it keep ourselves juiced up for frustration.

Most people seem to bring along a bag of recipes for frustration when they come to prayer. They come expecting to find Mr Fix-it. We expect that God is there to fix up all of our little needs. That he will pass our exams for us even if we did not study. That he will cure our great grandmother who is now 115. That we will win the lottery for us. (One man complained to God 'God, give me a break let me win.' God replied, 'You give me a break. At least buy a ticket!')

We come with a load of desires and we want him to fulfil them for us. By our prayers we want to control him, to manage him. But God will not be controlled. He will not be controlled by our goodness. We cannot force him to be, as we see it, good to us.

Likewise, it is well to note, we cannot control God by our badness. Sometimes people say that some happening is God's punishment for our bad deeds. This is very strong in the Philippines. But God has never given us control over him. He has not said that he will do bad because we do bad. On the contrary he has taught us that he will do good even if we do bad. Throughout the Old Testament he loved his people in spite of their infidelity and he gave us Christ to save us even in our sinfulness.

It is very foolish for us to set ourselves up as knowing what is good and bad, and then to expect God to follow us.

> The Emperor and the Prime Minister were out hunting one day when the Emperor broke his thumb. As the PM tried to bandage the thumb he remarked, 'Emperor, we never know what is good and what is bad for us.' In his pain the Emperor could see no point in such philosophising and angrily

pushed the PM into a deep well and left him there. As he proceeded on his journey he was captured by a cannibal tribe who were looking for a human sacrifice. They were just about to sacrifice him when they noticed the bandage and the broken thumb. They immediately released him as nothing blemished could be offered in sacrifice. At this the Emperor became very remorseful about the Prime Minister and went back to pull him out of the well, full of apologies. But the Prime Minister said, 'Why do you apologise? Can't you see that you saved my life!'

This matter was dealt with very directly by St James in chapter four of his epistle;

> Where do wars and battles between yourselves first start? Isn't it precisely in the desires fighting inside your own selves? You want something and you haven't got it; so you are prepared to kill. You have an ambition you cannot satisfy; so you fight to get your way by force. Why you don't have what you want is because you don't pray for it; when you do pray and don't get it, it is because you have not prayed properly, you have prayed for something to indulge your own desires.

We see then that prayer is not to fulfil our desires but to enable us to let go and transcend them. It is not to give us our wants but to enable us to transcend wanting. So often what we pray for is that our Ego's be pandered to. God loves us too much to answer such prayers.

St John tells us that 'Love drives out fear.' Yet, we pray mostly out of a sense of duty or of obligation. It was drummed into us in childhood that we had to pray, we had to go to church or if we did not we would go to hell. What a sad starting point for relating with God! We pray to placate a God up there. But how can we love someone that we are afraid of. When we are afraid we cannot grow because when we are afraid we are insecure, and our energies go into defending what is threatened inside us.

To pray we need freedom, freedom especially from fear and expectations. This is precisely why Christian meditation is so important. In meditation we have no words or images. Words or images – father, mother, king, slave – all have implications of servitude and fear. In meditation we just be, fully honest and true, basking in the sunshine of God's love.

When we meditate we repeat the mantra. When we start off we think that we will have this mastered in a week or two and we will be 'super' meditators. Soon we let go of this and all expectations. We find that we can't hold the simple little mantra. Do we get angry and give up in frustration? Or do we come to understand that prayer is a journey, a discipleship, a way. We do not pray to be successful. We pray to be faithful. As we let go of our ambitions and desires about the mantra we will find that we are letting go also of many of our other desires, dreams and plans for ourselves, others and God. We are getting out of the prayer 'trap' and becoming open to the God whose wisdom is beyond all of our wisdom and whose love is beyond anything we could hope for or imagine.

Prayer Beyond Past and Future: The Mass

The present moment is of the greatest importance in all traditions of meditation and not less so in the teachings of John Main. The only time that is real is the present moment. The past is gone the future has not come. The only time that we can be happy is now. The only time to be is now. How fed up God must be if we pass a field of flowers, a beautiful sunset, a work of art and do not even notice it! The present moment is a sacrament, a meeting of the secular and the sacred, the temporal and the eternal, the passing and the permanent. Every time we meditate we celebrate this sacrament. We celebrate a meeting that is at one and the same time a meeting with life and fullness, and also a meeting with death. Maranatha, Come Lord Jesus, is a moment of meeting with life. When we finally meet that life we are absorbed into it, we are then dead to this life. The Alpha meets the Omega.

If meditation is a mini-sacrament of the now, the Mass is the major one. When we come together to celebrate the Eucharist we tell the story of the last supper. We remember that story with such power that its elements become really present once more to us. We are present once more to Christ's offering of himself in a thanksgiving sacrifice to the Father in the Holy Spirit.

When we come to Mass we are called to set our daily chores aside and offer once more Christ's thanksgiving sacrifice to the heavenly Father. We listen to the story of how this Father has entered into history in the Old Testament through Creation, through Abraham and those whom he called, till finally he becomes one of us in his son Jesus Christ. In a homily we may be called to reflect on how that coming is happening in our midst today. We then offer bread and wine, solid and liquid, food and drink, symbols of all the dualities of our world. This offering, this total letting go of all, is then taken and consecrated. We invoke the Holy Spirit to come and change these elements of bread and wine so that they will become for us the Body and Blood of Jesus Christ. What we offered is transformed, transfigured, a metanoia takes place. We have the sacramental moment when what is temporary and passing meets the permanent and eternal. We have a moment of life and a moment of death. In the giving of life what is eternal goes into the temporal. At the moment of death what is temporal goes into the eternal. We have a union of opposites from which we can say:

> Through him, with him, in him,
> in the unity of the Holy Spirit
> all glory and honour is yours,
> almighty Father,
> for ever and ever. Amen

Then the species are returned to us, given for us, to bring about communion, a new unity with God, self and others in our day to day world.

The total giving that the Mass signifies was brought home to me very movingly some years ago. A contemporary of mine was due to get married when his bride-to-be was discovered to have cancer. She gave him the option to cancel the wedding but he insisted that they would go ahead as planned. Sixteen years and four children later I was at home when she was in her final illness. She had faded away to a shadow. In my conversations with her she was concerned only for the welfare of her husband and children after she died. One Thursday night I celebrated Mass at her bedside with her family. As I held up the host and said the words 'This is my body which is given up for you,' I saw her emaciated body in front of me. When she died next morning the words had new meaning for me.

In the Mass we have the major sacrament, being present and giving ourselves over in sacred signs, to the whole God, Father, Son and Holy Spirit in time and beyond time. Every time we meditate and every moment of our meditation we have a minor sacrament of that same presence and giving over. As we practice the minor sacrament we grow also in appreciation of the major one.

A Wider Vision of the Church

The Centralising Church

Supposing you were Christ on Ascension Thursday morning, what would you have been doing? The disciples had proved themselves to be a cowardly bunch of bunglers. Maybe if you or I were in his place we would have been drawing up memos or outlines or dictating instructions so that the work would run smoothly after our departure. But Christ did nothing like that. He left his memory and his Spirit in the care of the incompetent twelve. And it continues to be in our care today as his people, his Church.

The Church of apostolic times had a great appreciation of the presence of the Spirit of Christ. They broke the bread of the Word and the bread of the Eucharist together. They shared ministry and care for one another very unpretentiously. They were a body with many parts working together. It was a Church full of spirit and dedication. Human conflict and venality are also reflected in the accounts of their life. As the early Christians spread they were persecuted. Often persecution lead to their spreading further.

Then the Emperor Constantine in the fourth century made Christianity the religion of the empire. Was this a blessing or a curse? When it became easy to be a Christian it became hard to be one! This led Anthony the Abbot into the desert to live the Gospel values more radically and begin what came to be known as religious life.

As the official religion of the court, Christianity began to become institutionalised. The ministers of religion became court officials who needed to have clear duties to perform. They tended to appropriate all ministry to themselves. The clergy found a theology for themselves in the priesthood of the Old Testament which had been in fact abrogated by Christ. In the Old Testament the priest belonged to the Levitical family and was for that reason set apart. Jesus, though not a priest in the sociological sense, himself became the priest bridging the space between God and humanity. He showed that this bridging took place wherever there was genuine love. In the post-Constantinian Church, however, there was a new emphasis on the priest as a person set apart. We hear for the first time of clerical dress. It was condemned then as a reversion to a pagan practice! In the Old Testament the idea of the priest as mediator was very important. Special people were needed to mediate with God for the people. Jesus did not emphasise mediation. To know him was to know the Father. In the newly 'liberated' Church after Constantine we hear more of the need for a mediator with God. In this new Church the simple meal rite that Jesus left behind began to take on the esoteric trappings of the Old Testament sacrifices.

Then Arius came on the scene saying that Christ was not God. The Council of Nicea clarified that by defining clearly that the Son was one with the Godhead. So if the two were equal why separate them and risk being punished as a heretic. Theology tended to put God, including the Son and the Holy Spirit, more and more up in heaven. He was the authority 'up there' to be obeyed. So too his Church should be obeyed. The Church often spread with political expansion. It is axiomatic that every oppressor seeks a God to justify his oppression. A God that was 'up there and to be always obeyed' was very useful for colonisers. Often the colonisers served a civil and religious function. The mayor and the priest were often one and the same person. For them the way of relating to God would be the model of relating to civil and religious authority. If it was anathema to question God, then it would also be unthinkable to question the civil or religious authorities.

As the empire spread and languages changed the Church kept the one Latin language. A sign of unity, yes, but it created difficulties in communication with the masses of the faithful. The learned were left talking to the learned and the people were left out. Through the ages some free spirits read the Scriptures and recaptured the Spirit there. They challenged the institutional Church and said that some things it was doing (like selling indulgences) were not according to the original message and spirit. This made the authorities very uncomfortable. They condemned the reformers and excommunicated them and then they took away from the people the Book that had given them those revolutionary ideas.

During the ages movements that claimed to speak from the Spirit were discouraged or condemned, often rightly, but often because they were not neat enough and threatened the institution.

The industrial revolution created another problem for the Church. Up to then the rural feudal population felt secure around the church. The parish was the centre of the people's lives. As the people moved into the centres of population the clergy were losing their hold on them. This led consciously or unconsciously to a new emphasis on the sacraments. If the people had to come for the sacraments it made the clergy financially and psychologically more secure.

With the declaration of infallibility at the First Vatican Council the Church reached its pinnacle of isolation. It now declared ,in effect, that it had the truth and could define it with little or no need to consult the people of God. The triumphal institutional Church reached its peak in the reign of the princely Pope Pius XII.

It is interesting to note that devotion to Mary also reached its peak in the time of Pope Pius XII. In the 1950's we had the last great Marian Year and the definition of the Dogma of the Assumption. As the Church became more institutional, as God became more 'up there' in heaven, the healthy faith instinct of the people told them that God was close. As theology had put God far away the simple people kept him close through popular devotions, images, saints and particularly through closeness to Mary the merciful and caring mother of God.

But the Spirit has ever been with the Church. As the Church grows in its self understanding some aspects of it tend to grow out of proportion. The problem is seldom a problem of truth but rather that part of the truth has been taken and made into the whole truth.

With John XXIII and the Second Vatican Council the windows of the Church were reopened. The signs of the times were to be listened to. The Church was to be renewed and an essential part of that renewal would be contemplative renewal.

John Main emerged as a leader in that work by rediscovering the ancient Christian tradition of meditation and teaching it in a clear and relevant way for our times.

A Church that Reflects the Trinity

We often ask: 'What's in a name?' Actually there is a lot in a name. We very often buy a product because of its brand name. We believe or disbelieve a person because of associations with his or her name. So, too, the name we have for God is very important.

As we have seen already, if we call God Father or Mother we acknowledge our littleness before him and our dependence on him. Our God will be far from us in heaven above. Our theology will be based mainly on principles from out there. There will be a great emphasis on worship and we will have a lot of talking in our prayer. Our Church will be very institutional. We will have many devotions and things to do to keep his Church going. Our church spires will reach to the skies. Our Eucharist will be too holy to be touched by ordinary hands. The sacraments will be seen as the way in which the Church shares out the riches from above with us.

If our name for God is Son, our God will be seen as God-with-us, Emmanuel. Our theology will be more situational. We will start by listening to what the situation is telling us about God rather than by saying what God tells us about the situation. As all people are made in the image of God, he is read through listening to them and to the signs of the times. Our prayer will be a prayer of listening to God in Scripture and in reading the situation out of the mindset of the stories and values found in Scripture. Our place for God, our Church, will be very much in the community and our source of life will be in the activities of the community. Our church will be a prophetic church that is ever striving to fulfil it's option for the poor. God will be found and heard mostly in the midst of his people. We will be able to touch our Eucharist, the God who has become one of us and has made himself our food.

We will want to take the Eucharist to where the people are rather than bringing the people to the Eucharist.

If on the other hand our name for God is Spirit we will locate him mainly within our own selves and within the people of God. We will be conscious that we are his temples. He will be heard as he speaks to our hearts. We will pray by just being with him in silence, our Church will be a mystical Church. It will probably not be very well organised but it will have a lot of fervour and deep conviction about it.

Let us look at this wider vision schematically:

Name	Location	Prayer	Church
Father	Heaven	Worship	Institutional
	Above us	Talking	Sacramental
Son	Earth	Biblical	Communitarian
	Around us	Listening	Prophetic
Holy Spirit	Within us	Being	Mystical

In distinguishing these different ways of seeing God, prayer and Church, we are not setting one up against the other or condemning any one of them. All are valid and necessary just as the respiratory system, the circulatory system and the digestive systems are all essential to the body. There cannot be health in the whole body unless each system is working well. So too there is need for an efficient and balanced institutional Church. Records need to be kept, administrative decisions have to be made, there is need for central guidance and support. A vibrant communitarian Church is equally necessary. This is a Church that will stand with the poor and think and theologise from their perspective. A living mystical Church is equally necessary in order that there be a fullness of life and wisdom. To claim that any one of them is more important or more essential than the others would be to break up reality. To make part of the truth into the whole truth is, I repeat, a form of heresy.

However, I do believe that for centuries the Roman Catholic tradition has been fixated in the Father/Worship/Institution level.

My Protestant friends tell me that their traditions have been mainly stuck in the Son/Bible/ Good Works level. Both traditions apart maybe from the Quakers, the Society of Friends, have missed out on the Spirit /Being/Mystical level.

Today we are living in exciting times! Since the Second Vatican Council the Bible has been given back to the people in their own language. This is an event of phenomenal importance. It means that people can now theologise. They can read the Scriptures and read reality and interpret God anew from the Bible stories and from their own experience. And people today are better educated and more democratically inclined than ever before in history. While this is wonderful it also brings it's own fears and tensions.

To understand what is happening in the Church let us digress for a moment and see what is happening in families. In the past in most cultures wisdom came with age. The older ones were wiser by the very fact that they were older. Today, however, older people may be total illiterates in the world in which the young people live and move. Young people may be far in advance of their parents in many areas of knowledge. Yet they are not ahead in experience and wisdom. It is inevitable that there will be clashes between the two from time to time. This is inevitable and if handled in a mature and respectful way it can be very healthy. However, if parents will not listen to their children, to their world of experience and to their point of view, it is just as bad as if children will not respect their parents. Both can grow by respectful listening. But if either party is afraid that party will probably not listen to the other. If, however, there is true listening they will generally see that there is no longer need to be afraid.

This situation is exactly paralleled in the Church. The community has to respect the experience and the wisdom of the institution. Likewise the institution has to listen to the wisdom of the community. It cannot expect docile submission from an educated and thinking laity. It cannot set the clocks back.

The Basic Ecclesial Community is now accepted as the model of being Church in many parts of the world. This is a way of being church modelled on the early Christians where small local groups of people come together to break the Word of God and the Bread of the Eucharist and to work together to solve the problems of the local and wider community. It is a Church that is very conscious of the presence of the Holy Spirit dwelling in the hearts of each and energising and guiding each one in a very special way. It is a Church that leads to mysticism, to meditation. Indeed it is a Church that needs meditation very badly. Because without the detachment from the Ego that meditation brings there will not be the balance necessary for the listening relationship with the institutional Church.

A Wider Vision

When Moses and the Israelites were crossing the Red Sea the whole court of heaven was at a standstill, every angel was holding it's breath. When the seas closed in on the Egyptians there was a deafening applause. Then someone took a look at God the Father. He was crying, 'The Egyptians,' he said, 'were also my children!'

There is a lot for God to cry about in today's world. This generation has eaten up wantonly millions of years of creation. The forests, the fossil fuels, and the sources of clean water are being quickly consumed and destroyed. One part of the world progresses at the expense of the other. There is war within and between nations. Populations are exploding. How are they to be fed and housed and given a quality life? There is much of the world without any form of social security system and in the places where there is, it is often being exploited or creating its own form of dependency and poverty. The poor are imprisoned for petty crimes while the rich are considered clever when they get away with legal multi-million rip-offs. Women are under pressure to marry and have families and, at the same time, to remain independent and progress in their careers. Children grow up without parents to make them a home and find consolation and companionship and excitement in lifestyles that lead to ultimate despair. Everyone is rushing to make time and money to be happy but seem to have lost the ability to just BE.

But there are also so many things to be happy and hopeful about. There is a growing global concern about the environment and ecology. There is a new appreciation for the complementarity and equality between male and female. Many of the walls between East and West have been broken down. There is a new hunger and search for interiority and meaning. There is a new appreciation of the basic unity in all things great and small.

The hunger and search for interiority can be seen in the thousands that have been flocking to India, or to Eastern style Gurus in Europe and in the United States, looking for ways of stilling self and attaining wisdom and peace. It is estimated that 8 percent of all Americans are into some such activities. Australia has the highest rate of Yoga teachers per capita in the world. In the City of Perth in Western Australia there are 16 different kinds of meditation groups.

It is good that Christians are recognising the religious richness of the East but it is sad that in doing so many feel that they have to leave Christianity. It is estimated that 65 percent of the leadership in Transcendental Meditation are ex-Roman Catholics. It was part of the genius of John Main that he not only discovered the richness of the East, but rediscovered that same richness in the Christian tradition and taught it in a very simple way.

The movement toward unity is also quite remarkable. Recently a Muslim friend told me this story.

> Imam Alim heard the angels in conversation. One asked how many people had made the Hajj to Mecca this year and was answered that more than a million had gone there on pilgrimage. 'And how many,' asked the first angel, 'had pleased Allah?' 'Only the Hajj of the cobbler Ali,' came the reply.
>
> The Imam was puzzled and he went to Ali the cobbler and asked him if he were a Hajji, and if he had made the pilgrimage. 'Yes and no,' Ali replied. 'For twenty years my wife and I collected money to make the journey to Mecca. We were just ready to go a few months ago when my wife visited friends to say goodbye. She discovered that their child was sick and that they did not have the money for the operation that was needed. My wife and I decided to give them the money we had collected and we said, "We will let that be our Hajj."'

What a beautiful Christian story! It points to the fact that all religions at their base are beautiful and transcending but when they come to expression, the expression is often imperfect. The living out of religion has been compared to the tips of our fingers. The finger tips are very far apart but when we go back to their roots in our hands they are very close.

For the Hindu, as Swami Satyananda taught John Main, 'The Spirit of him who created the universe lives in our heart and in silence is loving to all.' For the Christian, the transcendent Father God is one with the immanent Spirit God.

We are at present on the verge of a new era in Christianity. Many of the institutions that we consider sacred are now collapsing and we must not lose trust that the Holy Spirit is in their collapse.

Vocations to the priesthood and religious life are falling off but biblical scholarship points out now that the priesthood and hierarchy as we know it began to emerge only about one hundred years after Christ. Some are appalled that confessions are dropping off but this way of dealing with sinfulness only became formalised in the ninth century. We place great importance on the Sunday Eucharist but Anthony the Abbot, the father of monasticism, was twenty years in the dessert without the Eucharist. There is very little that does not change. Faithfulness is not in clinging to a past which may be dead, but in recognising and nurturing God's lifegiving action in the here and now.

We have much talk today about priesthood, married priests and women priests. The problem here is about ministry rather than about priesthood. Because priesthood has become identified with power, prestige and possessions nobody wants to give over these worldly weapons to another. But priesthood is about ministry and ministry is about service and there is always a joy and a delight when others share service. A renewed Church will have to be one that focuses on service rather than on power. It will have to know how to let go of a lot so as to be free to answer the Spirit. It will have to be sensitive to all the wonderful realities that must be balanced in the glorious world that God gave us.

A truly renewed Church will be a Church that is at home with reality. According to the Christian stories that have created us, God is the centre of reality. So to be at home with reality is to be at home with the God of the Bible, a suffering God, a God who emptied himself taking the form of a servant, a God who is now indwelling in each of us. This will be a contemplative Church.

If the forest is to be green, the trees must be green. If the Church is to be contemplative, its members will need to be contemplative. That is why there is a great sign of hope for the Church in the many little groups of meditators that are coming together to just *be* in God's presence, in so many parts of the world. This is the first step towards the Church becoming what it claims it wants to be. This direction is expressed very beautifully in the Swiss Canons for the Mass, now approved for use in most parts of the world:

> Grant that in a world torn by strife
> Your Church may become an instrument
> in the service of unity.

> Make us attentive to all men and women
> that we may lovingly share in their sorrows
> and fears,
> in their expectations and their joys,
> and help them on the way to salvation.

> Put the right words on our lips
> each time we meet on our way
> lonely and discouraged brothers and sisters.
> Give us the courage to stand with our
> fellowmen
> whenever they suffer from poverty or oppression.
> Let your Church be a harbour of truth and freedom,
> of justice and peace,
> that she may be to each person
> a reason to keep hoping.

Meditation Centres

29 Camden Hill Road
London W8 7DX
England
Tel: 071 937 0014

Meditatio PO 552
Station NGD
Montreal H4A 3P9 Canada
Tel: 514 766-0475

24 Murray Road
Croydon
Melbourne
Victoria 3136
Australia
Tel: 03 725 2052

2745 Elson Avenue
Oakland
San Francisco
California 94602
USA
Tel: 415 482 5573

Holy Family Church
6 Chapel Road
Singapore 1542
Tel: 344 0046

Perpetual Help Church
Redemptorists, P.O. Box 139
Dumaguete City
6200 Philippines
Tel (63-3522) 52203

62 Park Avenue
Dublin 4
Ireland
Tel: 2693466

5F Chronicle Building
Pektite Ave/cor Meralco Ave
Pasig
Metro Manila
Philippines
Tel: 6333364